TL 589 ADV

2

0

Advances in Optronics and Avionics Technologies

WILEY – EC AERONAUTICS RESEARCH SERIES

Advances in Optronics and Avionics Technologies

Edited by

M. Garcia
European Commission, DGXII, Brussels

European Commission Aeronautics Research Series

JOHN WILEY
Chichester • New York • Brisbane • Toronto • Singapore

Publication no. EUR 15497 EN of the
European Commission,
Dissemination of Scientific and Technical Knowledge Unit,
Directorate-General Telecommunications, Information Market and
Exploitation of Research,
Luxembourg

Published in 1995 by John Wiley & Sons Ltd,
            Baffins Lane, Chichester,
            West Sussex PO19 1UD, England

            National   01243 779777
            International (+44) 1243 779777

Other Wiley Editorial Offices

John Wiley & Sons, Inc., 605 Third Avenue,
New York, NY 10158-0012, USA

Jacaranda Wiley Ltd, 33 Park Road, Milton,
Queensland 4064, Australia

John Wiley & Sons (Canada) Ltd, 22 Worcester Road,
Rexdale, Ontario M9W 1L1, Canada

John Wiley & Sons (SEA) Pte Ltd, 37 Jalan Pemimpin #05-04,
Block B, Union Industrial Building, Singapore 2057

British Library Cataloguing in Publication Data

A catalogue record for this book is available from the British Library

ISBN 0 471 95362 8

Typeset in 10/12pt Palatino by Keytec Typesetting Ltd, Bridport, Dorset
Printed and bound in Great Britain by Bookcraft (Bath) Ltd
This book is printed on acid-free paper responsibly manufactured from sustainable forestation,
for which at least two trees are planted for each one used for paper production.

Contents

2   New optical sensor concept for aeronautics (NOSCA)   65
M. Turpin

3 Integrated modular avionics general executive software (IMAGES)   105

D. Graves, R. Meunier and P. Aldegheri

Foreword

Over the last thirty years the European aeronautical industry has achieved a respected and internationally successful position. Aeronautical products made in Europe are able to secure large market shares, and in some instances have even become dominant. Nevertheless, the level of global competition continues to be set by the United States of America, which is traditionally committed to pre-eminence in this field.

Despite the importance of the European dimension, it is only in the comparatively recent past that the European Community has started to play a significant role in the technological challenge in aeronautics.

In 1988, the major European aircraft manufacturers presented the EURO-MART study (European Cooperative Measures for Aeronautical Research and Technology) to the European Commission. The study identified areas of research which were considered to be critical to the future competitiveness of the industry in world markets. Separate reports, conveying views on the content of a European research activity in aeronautics, were submitted by representative groups of the European Aero-Engine Manufacturers and the European Aerospace Equipment Manufacturers and Systems Suppliers.

Following these extensive consultations with industry, the Commission launched its first dedicated aeronautical research programme in March 1989. This initiative became known as the aeronautics 'pilot phase' and was in fact Area 5: Specific Activities Relating to Aeronautics, of the BRITE/EURAM-Programme, itself a Specific Programme within the Community Second Framework Programme for Research and Technology Development (1989–1992).

This action is being continued under the 3rd EC Framework Programme (1991–1994) and will be pursued under the 4th EC Framework Programme (1994–1998).

The pilot phase comprised 28 projects, covering four technology areas: aerodynamics, acoustics, airborne systems and equipment and propulsion systems.

Despite a relatively small budget 35 MECU over two years, this exploratory action achieved considerable success in stimulating wide-ranging co-operation between all types of actors. These included airframe, engine and equipment manufacturers, small- and medium-sized enterprises, research centres and universities from the Member States of the Community and also from EFTA countries.

This EC Aeronautics Research Series provides the opportunity to present the results of those research projects which have reached completion, and to illustrate the achievements of this pilot action.

This volume includes the three projects which, in this pilot phase, carried out work in the area of airborne systems and equipment technology.

In the optical data transmission (ODT) project six aircraft manufacturers together with equipment companies, research centres and universities undertook a comprehensive effort to evaluate the optical approach for data transmission onboard aircraft. Owing to their decentralised systems future aircraft will require high integrity, high speed and above all electro-magnetically passive data transmission systems. The project included development of fibre optic data links, system topologies and the assessment of the appropriate power budget together with simulations, laboratory tests and flight demonstration of fibre optic networks.

The aim of the complementary NOSCA project was to develop optical fibre sensors for the measurement of pressure and acceleration. Equipment manufacturers worked together with research centres and universities in this project. The sensors were defined, designed, assembled and successfully tested in the laboratory environment. In terms of networking, the coherence multiplexing technique was implemented and the development of a specific processing technique (soft- and hardware aspects) permitted the recovery of multiplexed information from different sensors without drastic crosstalk effects.

In the IMAGES project, airframe manufacturers together with equipment companies, including small- and medium-sized enterprises, research centres and universities dealt with the definition of the functionalities of avionic-oriented executive software in the context of integrated modular avionics, and with the usability of ADA language for avionic applications. With regard to the executive software all services were defined, notably the real-time executive features, communication management, initialization, interrupt and execution handling and the avionics components library. This and the results of the ADA studies enabled integration in the ARINC standardization.

The notable technical successes achieved in these projects illustrate the added value of collaboration at European level.

Thanks are due to all partners of the three consortia, particularly to the scientists, engineers and managers who contributed to the success of these aeronautical projects, reflecting the high level of collaboration attained under a common interest of advancing the European Union technology base in this crucial sector.

Thanks are especially due to the authors of the three technical project reports.

M. Garcia
Brussels
October 1994

1 *Optical data transmission in aircraft*

R. Bogenberger*

This report, for the period February 1990 to January 1992, covers the activities carried out under the BRITE/EURAM Area 5: 'Aeronautics' Research Contract No. AERO-CT89-0012 (Project: AERO-P1062) between the Commission of the European Communities and the following:

Deutsche Aerospace* (coordinator), Germany
Aerospatiale, France
Alenia, Italy
British Aerospace Airbus, United Kingdom
Cert, France
Dassault Aviation, France
Farran Technology, Ireland
Fokker, Netherlands
Hellenic Aerospace, Greece
NLR, Netherlands
NTU Athens, Greece
Smiths Industries, United Kingdom
STC, United Kingdom

Contact: Mr. R. Bogenberger, Deutsche Aerospace AG, Aircraft Group, Dptm. LME 362, Postfach 801160, D-81663 München, Germany (Tel: +49/89/6072.6219. Fax: +49/89/6072.9445)

0 Abstract

To promote optical data transmission in avionics, the European Commission launched the project 'Optical Data Transmission (ODT)' in spring 1990. The ODT project will cause significant progress in fibre optic (FO) data transmission techniques for aircraft within the context of the Brite/Euram program.

The experimental study 'Optical Data Transmission (ODT)' was carried out by a consortium of 13 European Aircraft/Electronic Industrial Entities, Research Institutes and Universities within the Brite/Euram program, to demonstrate the feasibility of fibre optic technology for use in future onboard data transmission networks.

The project is divided in seven tasks and each task is again divided into a number of subtasks. The investigations and results of tasks and subtasks are described in this final report.

This report will resume results of this study. In particular investigations with respect to system requirements and the definition of data transmission system architecture will be discussed. A synopsis of fibre optic standardization activities and outlines of a first generation fibre optic data transmission system will be presented. Fibre optic data links, various system topologies and the appropriate power budget, as well as results of experimental investigations, e.g. simulations, test methods, laboratory evaluation and flight demonstration of fibre optic network, are described.

The goal of the ODT project is to define and evaluate concepts for optical data transmission systems for application in future civil aircraft.

- The specific requirements for the integration of advanced sensors, signal processors, decentralized computers and video terminals will be determined.

- Fibre optic architectures (topologies) will be reviewed, designed and assessed to meet EMI requirements (e.g. lightning and electromagnetic disturbances) and to permit reliable transmission of data at rates significantly higher than in present aircraft.

  Specific objectives in this part are: to arrive at agreed ODT designs with future expansion potential; to identify in co-operation with suppliers the present shortcomings of optical components/modules; to establish draft standards for high speed optical data transmission systems to facilitate cost reduction; to develop and define test methods with the aim to eliminate unscheduled maintenance and thereby to reduce life cycle costs.

- Laboratory and flight tests on ODT demonstration systems will be carred out to identify areas for further design and development of components/ modules.

To define the requirements for data transmission systems, the data flow and exchange between the different units (such as sensors, signal processors, decentralized computers, display units and video terminals) was analysed and reported with respect to the number of subscribers, data transmission rates, maximum message length and repetition of messages. In order to establish a solid basis for the design of an advanced avionic system, the expected limiting conditions for future aircraft systems have been defined.

Fibre optic system topologies (linear, ring, mesh and star topology) were worked out in the first period of the project. The advantages and disadvantages of active and passive star (transmissive as well as reflective) and the coupled networks were compared with respect to the number of subscribers, reliability and access methods.

Applicable protocols, access methods (e.g. token passing, command response, CSMA/CD, CSMA/CA) and addressing of databus subscribers were shown, with respect to both the characteristics of the expected data traffic and possible limitations arising from the use of optical transmission instead of electrical transmission.

Growth potential, protocol, access method for system subscribers and optical power budgets will lead to the selection of topologies suited for ODT network implementation and for networks which may be composed of several subsystems. The selected protocol and topology are the subject of current simulations.

After the selection of specific systems topologies and layouts, the requirements for the modules will be established. The selection of the first generation of ODT system will be performed on the basis of availability, reliability, installation and maintenance conditions and cost, etc. Test methods have been identified and worked out for ODT systems and ODT system components.

Laboratory and flight demonstrations showed that information transfer in ARINC 629 format and as well as in high-speed token bus (HSDB) format via a fibre optic bus system is feasible. Thus, a demonstration of data transmission via a fibre optic network, capable of simulating concurrent best and worst case losses, has been carried out using available hardware, slightly modified.

During test flights aboard a Fokker 100, relevant bus parameters have been monitored and recorded using existing flight-test equipment and predefined test networks. In parallel the compliance with civil aircraft environmental conditions has been monitored. The experiences during installation and flight testing are described in this final report.

Two data transmission systems have emerged as the result of the ODT study. In the first stage areas that require standardization have been identified. The aim is to finalize the specification of ODT system/components with due consideration of the topology. These activities will take into account European and international standardization organisations.

This joint effort is aimed at developing the technology for advanced avionic ODT systems in Europe. This will be achieved by the co-operation of the European partners who are technology leaders in aircraft data transmission systems such as STANAG 3910, Optical STANAG 3838, ARINC 629 and ARINC 429.

1   Introduction

The use of data transmission via optical media in the area of telecommunications is rising steadily. The driving factor in this case is the very high transmission bandwidth, combined with a comparatively small transmission loss.

The use of fibre optic techniques in aircraft and space vehicles opens up a great variety of applications, caused by enhanced efficiency (e.g. data rate) on the one hand and higher reliability and resistance against electromagnetic interference (EMI, lightning) on the other hand. The large number of different systems in an aircraft lead to a variety of requirements on data distribution networks.

Considering flight control systems, for example, special attention is on reliability and robustness against electromagnetic interference. Looking at the cockpit area, especially to the links between display and display computer, the high data rate is an important point of view. Networks located in the actuator or engine area require high reliability and live time in harsh environment, e.g. high temperature and vibration load. Links used in the sensor area of avionics need to be designed for an extremely high data transmission rate. The introduction of optical data transmission requires new advanced methods in installation and repair of fibre optic cables, connectors and components, accompanied by novel test procedures to accommodate the fibre optic technology.

The well-known principle of transmitting data between sources and destinations via point to point links will be replaced more and more by the method of data propagation via buses. The establishment of digital computers in modern aircraft represents a major step towards the utilization of digital buses.

Assessment of fibre optic systems: in regard to the aircraft applications, efforts to characterize the performance limits of data transmission system should be made. Bus systems can be characterized by data transfer capacity of network, station separation and the length of harness and the maximum applicable number of subscribers; these factors are of importance in regard to data transfer reliability (in terms of signal to noise ratio and transmitter output power in copper or in fibre optic networks).

Requirements such as high reliability and extreme real-time demands for aircraft systems lead to a new era of transfer protocols. Now they can be mostly made based on computer/communication techniques. In addition the benefits derived from protocol, coding and access mechanism should be investigated.

2   Research objectives

2.1   Task 1: investigation of the requirements for aircraft optical data transmission

To define the requirements for data transmission systems, the data flow and exchange between the different units, such as sensors, computers, displays

..., was analysed with respect to the number of subscribers, data transmission rates, maximum message length and repetition of messages.

In order to establish a solid basis for the design of an advanced avionic system, the expected limiting conditions for future aircraft systems have been defined.

2.2 Task 2: review and assessment of system architectures, components and procedures

Fibre optic system topologies (linear, ring, mesh and star topology) were discussed. The advantages and disadvantages of active and passive star (transmissive as well as reflective) and the T-coupled networks were compared, with respect to the number of subscribers, reliability and access methods. Active and passive components/modules for the various system topologies were reviewed. The assessment will cover the performance and will present specific advantages and shortcomings; it will focus on desired improvements. Applicable protocols, access methods (e.g. token passing, CSMA/CA) and addressing of databus subscribers was shown, both with respect to the characteristics of the expected data traffic and possible limitations arising from using optical transmission instead of electrical data transmission. Growth potential, protocol, access method for system subscribers, optical power budgets will lead to the selection of topologies suited to ODT networks which may be composed of several subsystems. The selected protocols and topologies has been the subject of simulation.

2.3 Task 3: establish implementation guidelines for bus systems

This task gives a preliminary set of guidelines for the application of ODT in aircraft. The task is divided into three subtasks, as follows:

- system design guidelines

- manufacturing installation guidelines

- component selection guidelines

2.4 Task 4: preliminary specification and selection of first generation of ODT systems

This task focuses on a first generation of ODT system for use in future aircraft. The task is divided in two subtasks and in a common introductory note as follows.

- Introductory notes to:

  system requirements (2.4.2.2.1)
  selected topologies
  power budget
  system specification

- Preliminary specification of components and modules
  After the selection of specific system topologies and layouts, the requirements for the components, modules and relevant interfaces has been
  established.

- Selection of first generation of ODT systems.
  On the basis of availability, reliability, installation and maintenance conditions and costs etc. fibre optic components and system architecture have
  been selected.

2.5   Task 5: development and definition of test methods

This task is to identify and report on the various methods of optical measurements available for optical data transmission, e.g. acceptance tests, methods
for testing complete fibre optic systems installed or self test methods.
  This task includes 5 subtasks, as follows:

- identifications of methods to measure, monitor and detect faults and deterioration of performance

- acceptance test methods of the fibre optic harness

- acceptance test methods for module interfaces

- definition of test methods for complete fibre optic system installed in A/C

- definition of system self test requirements for performance deteriorations
  and endurance monitoring

2.6   Task 6: experimental activities: laboratory and flight demonstration

2.6.1   Laboratory demonstrations

A laboratory demonstration showed that the information transfer in ARINC
629 format and as well as in high-speed token bus (HSDB) format via a fibre
optic bus system will be feasible. Thus, a demonstration of the transfer of data
via a fibre optic network, capable of simulating concurrent best and worst case
losses, has been carried out using the available hardware, slightly modified.
  The following parameters will be measured and the results will be evaluated:

system performance, access methods, system delays, path loss and dynamic range of testbench hardware.

2.6.2 Flight test demonstration

During test flights, relevant bus parameters has been monitored and recorded using existing flight-test equipment and predefined test networks. In parallel the compliance with civil aircraft environmental conditions has been monitored. The experiences during installation and flight testing are described in the final report.

2.7 Task 7: preparation of draft specifications

The first stage of investigation in this task was to identify areas which requires standardization. The second stage deals with standardization bodies and existing/current standards. The third stage shows examples of existing standards/specification applicable to ODT. These activities will take into account European and international standardization organizations.

3 Research activities and results[1]

3.1 Task 1: investigation of the requirements for aircraft optical data transmission

3.1.1 Subtask 1.1: investigation of the required and expected data transmission

Responsible partners: Aerospatiale, AIT, BAe, CERT, MBB

Activities In order to define the requirements for data transmission systems, the data flow and exchange between the different units, such as sensors (e.g. communication, air data, primary/secondary flight control system, electrical and hydraulic control system, radar), signal processors, decentralized computers, display units and video terminals, will be analysed with respect to the number of subscribers, data transmission rates, maximum message length and repetition of messages. Here, the experience of MBB, Aerospatiale and BAe will be utilized, firstly to obtain the characteristics of present data transmission

---

[1]The work in the project ODT was organized in seven approximately stand-alone tasks; even more subtasks inside the task are more or less independent of each other. This prerequisite leads to individual final task reports with individual descriptions of activities followed by the dedicated results and conclusion of each task. It seems to be much better for clearness to stay at the original structure of the project. In the following chapters the research activites, the results and at least some specific conclusions will be described in Chapter 3, followed by a task number derived from the ODT project structure.

systems, and secondly to extrapolate the requirements for future aircraft. Finally, the basic requirements will be agreed between the partners.

Subsystems involving the main flight functions (e.g. primary/secondary flight control, engine, fuel and landing gear systems) or high data flows (e.g. navigation and display unit systems) are analysed. The characteristics of other subsystems have little influence on global system architecture and are not analysed.

Results In order to evaluate the work presented in this subtask, it is necessary to compare the different communication networks that have been developed recently (e.g. A320 and current Boeing aircraft) and new aircraft currently under development (e.g. A330 and A340). The Airbus Industries A320 (medium sized, twin engined commercial airliner) and A340 (large sized, four engined commercial airliner) also show the different sizes of communication network that are required for different sizes and types of aircraft.

The following list summarizes the problems and trends currently being faced by aircraft manufacturers. The total volume of information transfer (messages) is increasing (analogue and digital) which in turn increases the number of electrical links needed and therefore the weight (and the transfer of discrete signals onto data buses will also increase the number of digital links). The number of electrical components is increasing with the development of newer and larger aircraft (longer links, introduction of new systems, etc.) which makes it more difficult to achieve adequate EMC. With the development of similarly sized aircraft, the number of computers and controllers has increased with each new generation. The memory required for digital electronic processing is increasing with aircraft generation. Computation power is also rapidly increasing with time. EMI and lightning protection requirements for electrical systems are costly and can impair the performance of the system.

For the purposes of this summary, a comparison will be made between the ARINC 429 databus currently used by all Airbus Industries aircraft, and ARINC 629, which is an example of the next generation of data buses; it is considered to be the prime candidate for adoption for future Airbus aircraft, because it is considered to have the best performance, is the furthest defined and is closest to standardization.

A full analysis of topologies and protocols will be carried out in Task 2, and it should not be taken for granted, before the results of the studies to be carried out in the following tasks are known, that ARINC 629 will be the protocol chosen for ODT (in Task 2 technical and economical aspects (such as life cycle costs) are assessed), although it is most likely to be chosen. Problems are specific to the nature of the databus presently used.

All present Airbus aircraft electrical systems are based upon the ARINC 429 databus, as the means of digital communication between all aircraft systems.

Architecture Simply, ARINC 429 is a databus that employs point to point topology. ARINC 429 is a bus with a single transmitter and multiple receivers.

All data are electrically transmitted over a single twisted and shielded wire pair in one direction only. There are two levels of data communication. For a non-critical system, a transmitting subscriber may 'talk only' to a maximum of 20 subscribers on a twisted wire pair. No acknowledgement is required by the receiving subscriber to confirm that it has received the message. For a critical system, where large amounts of verified data are to be transferred between subscribers, an acknowledgement is required to check that all data have been received. This requires the use of two pairs of shielded twisted wires instead of one and therefore increases the weight of the system.

By contrast, ARINC 629 is a real-time communication system characterized by its linear bus topology. It is autonomous in nature and assures equal access to each participating subscriber. The system consists of a physical bus and stubs connecting the bus to each subsystem. A data source or sink is connected to the global bus by means of a bus coupler.

ARINC 629 has two classes of data service: commands/status/requests for information are treated as non-acknowledged messages; directed messages/file transfers which are globally significant messages are required to be acknowledged.

With the development of newer and larger aircraft and the expansion in the number of electrical system subscribers that this involves, there will be a larger weight cost for the future introduction of an electrical data communication network. Direct comparison of a 200/280 μm tight buffered optical cable as specified for EFA (weight approximately 4 g/m) and an electrical data transmission 2 × 24 gauge twisted optimized insulation cable as used on A320 (weight approximately 10 g/m) gives a weight saving of over 50%. This does not include additional savings permitted by the high bandwidth capability of an optical fibre. This could allow one optical fibre to replace several electrical cables and provide growth potential. It is considered that even installing redundant cables (for repair, etc.) will allow large weight savings.

Characteristics ARINC 429 electrical cables requires shielding at each connection and termination. All data transmitted conform to the bi-polar return to zero format and can be affected by electro-magnetic interference.

ARINC 629 electrical cables will be unshielded twisted pairs (current mode) with possibly shielded stubs of up to 40 m in length connecting subscribers to the databus. ARINC 629 is a multiple transmit and receive databus, compared with 429 which is a single transmit and multiple receive network. A single 629 link can replace several 429 links and perform the same operations. A fibre optic network will have an advantage over electrical networks because grounding will not be required. Each subscriber (and active components) will still require grounding.

The highest data transmission rate allowed on an ARINC 429 link is 100 kbits/s. The required data flow is increasing by a factor of ten, in time, with each new electrical databus standard (e.g. ARINC 429 in 1977 and 629 in 1990). By extrapolation of the rate of progression in the required data rate, it is probable that tomorrow's aircraft will need a 10 Mbits/s data rate within ten years or even sooner (and is the data rate agreed by the partners for the basis of the analysis of topologies in Task 2.1). Beyond the current development stages, data rates of up to 100 Mbits/s should be achievable (as the telecommunications industry is proving with FDDI/X3T9.5).

Copper cables are capable of reaching the 10 Mbits/sec data rate, but at a cost. The copper must be of a high purity, and EMC protection improvements are required. These two factors contribute to making the use of copper for ARINC 629 very expensive compared with previous standards, and make the use of optical fibre technology more attractive.

As advances in technology continue, the higher data rates and longer overall message lengths will eventually become too large for the smaller bandwidth of the electrical (copper-based) databus for future aircraft applications, compared with fibre optics. The bandwidth of optical fibres is great enough to allow a wide scope for expansion of the volume of information that future data transfer rates would require. The signal bandwidth of copper cables is of the order of MHz for the shorter lengths (compared with telecommunications) of an airplane, whereas that for optical fibre is of the order of GHz (or greater). This enables far more information to be transferred through optical cable than electrical cable and allows one optical fibre to replace several electrical wires, simplifying cabling requirements for aircraft systems.

An optical ARINC 629 should show marked weight advantages over the electrical ARINC 429 due to the reduced cabling requirements of the (electrical) databus, combined with the advantages of reduced cabling for optics, compared with electrics.

Message lengths   ARINC 429 defines the size of the data words to 32 bits, with a word format as follows:

label	8 bits
parity	1 bit
sign/status matrix	3 bits (binary format)
source destination indicator	2 bits + 13/14 bits

When the + sign is considered part of the data, one word may have 19 or 20 bits of data. Sometimes the SDI is used for data as well, giving a maximum of 21 bits of data.

The maximum number of 32 bit words that can be accepted by any receiving subscriber in a single message is 255. ARINC 429 defines a bit-oriented file transfer protocol. A file consists of link data units (LDUs) and each LDU consists of words. The maximum file size consists of 255 LDUs, with 255 ARINC words (not all data) in each LDU. (The actual amount of data for each file would be up to 1.3 Mbits, assuming 20 bits of data per word. Some of the words may not contain any data, e.g. destination codes.)

ARINC 629 for future aircraft will require data words of 20 bits (of which 16 contain data information). The maximum word string length is 257 words (aperiodic message), or 5140 bits in total (80% containing data and 20% containing identification and priority bits). The maximum length of a data transmission message for a terminal (subscriber) will be 31 word strings (periodic message), or 159 340 bits. The rates at which labels are updated depend on the importance (priority) of the parameters being monitored. For example, in the navigation system (ref. 1.1.8), a sample rate of 10 ms is common, whereas the fuel system (ref. 1.1.4) may have sample rates of up to 1 second.

Maximum number of subscribers   On current Airbus aircraft, the use of ARINC 429 limits the number of subscribers in any system or subsystem to 21 (1 transmitter and 20 receivers). From this subtask, some of the systems have a number of subscribers which are close to the limit set by ARINC 429 (e.g. navigation system has 19 computers). Future aircraft systems (navigation, communications, etc) will increase the maximum number of subscribers to more than 20. Therefore a databus with a preferably unlimited maximum limit on the number of subscribers is desirable. Such a databus will allow continuous expansion of the number of subscribers as required for each generation of system/subsystem.

ARINC 629 will allow an increase of subscribers (electrically) from 20 to at least 120. Implementing this electrically provides the possibility for 120 or more subscribers, with more than one subsystem sharing a single data terminal. When actuators/pumps/sensors/etc. become smart these could be considered 'subscribers': i.e. they could feasibly be hooked directly to the databus. Considering one A320 system, such as landing gear, this could add 32 or so sensors to the databus system via an interface.

Future data flow projections   The projected data flow for a future bus system depends upon three factors: equipment integration, architecture with a multiple access bus, and new systems.

Equipment integration   To reduce data flow (currently the objective of electrical bus systems) requires a full integration of equipment with a high level of interconnection.

Architecture with a multiple access bus   The organization of the systems will be influenced by the concept chosen on the multiple access bus. On the one hand there could be centralized architecture with all equipment connected to the bus, and on the other hand a completely decentralised architecture with multiple buses.

New systems   Future systems that may be revised or added are: navigation, traffic surveillance (via satellite), airline operations (via satellite), replacement of mechanical/semiconductor switching with optical switching, audio systems, software loading, and cabin management.

It is not possible to estimate exactly what the future data transmission flow requirements will be without knowledge of the topology, protocols, etc., (which will be addressed in Task 2) as applied to the specific application. Therefore, for aircraft applications, the best guide to the future data flow can be obtained from projecting the rate of progression from past to present systems into the future.

For an aircraft it is difficult to determine the data flow for all the systems present. It is calculated on the basis of the exact flow for each data link at one instant in time, summed. This will vary depending on which moment in time the data flow is calculated. Therefore a maximum value possible for the aircraft is the only way of indicating the probable evolution of data flow for future aircraft.

Reliability   For present Airbus aircraft (ref. ABD 0048), the memory MTBF (mean time between failure) shall not be less than 40 000 flight hours for any computer. Referring to section 1.2.4 'Reliability and Repair', it is possible to quote the estimated required life expectancies for aircraft optical components, based on the MIL-HDBK-217E that the equivalent electrical components must satisfy.

Singleway FO connector (e.g. HA)	10^5 hours
Splice	10^7 hours
Cable	10^{10} hours per metre
Couplers	10^7 hours
Active components (LEDs, etc.)	1.54×10^7 hours at 25 °C
	5.73×10^4 hours at 125 °C

System life expectancies are covered in detail in Task 2. An example (from 2.1.3), of a 27 subscriber star layered topology network has an MTBF of approximately 1.25×10^4 hours.

Latency   Latency is defined as the time from the instant that an amount of information needs to be sent from one subscriber to another until the instant that the complete message is received at the destination. It is most critical for the transfer of information between the navigation and flight control systems. Current aircraft using ARINC 429 have message latencies of approximately 2 ms. The exact latency will depend upon the architecture, the distance between subscribers and message size, as well as the rate of transmission. ARINC 629 will reduce the latency to an average of approximately 0.5 ms (which can be a minimum of 20 μs, depending on the distance between the subscribers) for the transmission of a single word between two subscribers.

For a system, an example with the following assumptions provides a latency estimate:

32 subscribers
each subscriber transmits messages of 20 words of 20 bits (message transmission time = 0.2 ms)
each subscriber transmits once in a cycle

The average bus access delay would be 16 (half the number of subscribers) × 0.2 ms (time a subscriber holds the bus) = 8 ms. So the average message latency would be: 8 + 0.2 = 8.2 ms, excluding propagation delay and queueing time.

Identification of critical systems From the contents of this subtask it is possible to identify these systems: flight controls, engine controls, fuel system, display units, and navigation.

One way to estimate the maximum potential data transmission rate for an aircraft (assuming that only ARINC 429 links exist, to keep the calculation simple) is obtained by multiplying the maximum possible transmission on one link (100 kbits/s) by the total number of links. This gives the maximum potential data transmission for ARINC 429, e.g. maximum data rate (100 kbits/s) × 100 links = 10 Mbits/s. For a more precise estimate of data flow, it is possible to take an example of one aircraft system, and assume that the proportion of high speed to low speed ARINC 429 links applies to the whole aircraft. In this case the primary flight control system is examined for A320 and A340, using the data contained in Section 1.1.1. From this the following information can be determined.

For the A320 PFCS, 22% of the digital links are HS and 78% LS. For the A340

Table 1   Summary of Subtask 1.1 (A320) systems

System	Computers	Refresh rates (ms)	Data flow (kbits/s)
1.1.1 Primary flight controls	7	10–120	172
1.1.2 Secondary flight controls	2	4–20	8
1.1.3 Engine controls	1	10–50	19
1.1.4 Fuel	2	20–4000	0.960
1.1.5 Landing gear	4	20–4000	3
1.1.6 Pneumatics	4		1
1.1.7 Ice and rain protection	5		0.5
1.1.8 Navigation	19	10–1000	640
1.1.9 Display units	13		1072

Table 2 Data flow

	A 310	A 320	A 340
Volume of AFCS system components (example) (litres)	134	69	31
Total volume of digital electronics (litres)	745	760	830
Number of digital components	77	102	115
Volume of on-board software memory (Mbyte)	4	10	20
Number of digital links	136	253	368
On-board calculation power (Mips)	60	150	250

PFCS, 45% of the digital links are HS and 55% LS. It is assumed that all A310 links are low-speed ones.

Using the total number of digital links stated in Table 2 and applying the above percentages gives the following result, combined with the maximum potential data flow available (growth potential) for ARINC 429:

	A310	A320	A340
Maximum potential volume of data flow (Mbits/s)	13.6	25.3	36.8
Estimated actual volume of data flow (Mbits/s)	1.7	8.0	19.1

Generally, less than 62.5% of any word will be data bits transferring information between subscribers, the rest being parameter label, destination code and priority bits, etc.

When ARINC 429 specifies a data transmission rate of 100 kbits/s, the number specified is the data transmission rate on one link only. To estimate the data rate for a system, it would be necessary to measure the individual data bits being transmitted on each link within the system, at an instant in time. The value of the data transmission rate will vary, depending on the level of data traffic on the links at the time in question. Therefore one way to represent the level of data transfer for a system would be to consider the maximum number of bits which can be transmitted per second for the system. This can be approximated to the sum of the data transmission rate (assuming full data transfer capacity is used) multiplied by the number of links in the system.

This maximum data volume will not be carried on one databus but on several dedicated system buses (flight control, navigation, utility, etc.)

The required bus bandwidth (Baud rate) is not identical with the nominal data rate (bit rate), but is much higher. Line coding for better integrity of transmission increases the bandwidth, e.g. by 20% in the FDDI 4/5 coding scheme, by a factor of two in Manchester Bi-phase encoding. The nominal transmission capacity cannot be utilized by 100% due to collision problems; depending on the access mechanism safety (= time) critical applications may only allow a maximum bus load of 20 to 30% to ensure timely data delivery.

From these facts it may be concluded that ARINC 629 may already be a bottleneck for the next generation of aircraft system design.

The information about data flow has to be taken with care because the method of analysis is not identical for each system: redundancies are sometimes included and sometimes not included, and differences between considered aircraft (i.e. sharing of the functions between systems) are not identified.

Conclusion   Finally, it is important to realize that the requirements for data transmission in aircraft cannot simply be deduced from present avionic system demand. This allows only for moderate enhancement of performance but not for really novel solutions. Development always is an interaction between technology offer and systems demand. In this sense existing and upcoming aircraft systems do not seem to need data transmission rates in excess of 5 to 10 Mbits/s, but currently unused concepts like Electronic Library System (e.g. onboard maintenance handbooks), discussed for Airbus A330/340 requiring about 40 Mbits/s, passenger information and entertainment systems with individual displays requiring more than 100 Mbits/s or 'smart structures' with enormous amounts of sensor data can only be realized with high-capacity data storage and transmission. Moreover, novel and as yet unknown ideas are likely to be stimulated by high-bandwidth/capacity data processing systems.

The implementation of ODT in civil aircraft may profit from a true technological spin-off from military developments. Because of their higher data processing requirements, military aircraft of the next generation will already have fast optical data buses (20 Mbits/s), and so optical technology also suitable for civil aircraft will be on the market within the next few years.

3.1.2   Subtask 1.2 system requirements and limiting conditions

Responsible partners: Aerospatiale, AIT, AMD, BAe, CERT, Fokker, MBB, Smiths

Activities   In order to establish a solid basis for the design of an advanced avionic system, the expected limiting conditions for future aircraft systems must be defined. A detailed study will be performed by all the partners to establish the current and future trends for system requirements such as: environment conditions (temperature range and cycling, humidity, shock and vibration levels, etc.), electromagnetic compatibility (EMC), installation and maintenance, reliability and repair, and growth potential for future extension.

After a general study of environmental conditions, some typical aircraft environments have been considered in order to cover all the commercial aircraft types. Because of the difference between optical and electrical technology against electromagnetic problems, a specific study has been made of this subject. A preliminary approach on maintenance and repair limitations has been carried out. Some investigations about expected future environmental conditions have shown that nobody knows exactly today what the future will bring.

Results   In order to evaluate the data flow for the future bus it is necessary to make a projection on the future at three levels: equipment integration, architecture with a multiple access bus, and new systems.

Equipment integration   A reduction in the data flow could only be obtained by a full integration of equipment with a high level of interconnection. On one side flight management and guidance, display management, aircraft systems monitoring (including warning), and maintenance and fault isolation. On the other side flight control and isolation.

Architecture with a multiple access bus   The organization of the systems can be largely influenced by the concept chosen on a multiple access bus. On the one hand there is a centralized architecture with all equipment connected to the bus. On the other hand a completely decentralized architecture can be envisaged.

New systems   It is important to list the future systems that may be candidates. Navigation, traffic surveillance (satellite data link), airlines operation (satellite), aircraft systems (electrical remote switching . . .), audio (Interphone . . .), software updating (data loading), cabin management . . . The evaluation of data flow for future systems is relatively difficult and will depend on architecture.

Evolution of digital systems   For identical functions, the volume of digital electronics is divided by two every five years. For example, volumes of AFCS:

A310 = 134 litres   A320 = 69 litres   A340 = 31 litres

The total volume of digital electronics has a tendency to increase with time:

A310 = 745 litres   A320 = 760 litres   A340 = 830 litres

The number of digital equipment has increased:

A310 = 77 litres   A320 = 102 litres   A340 = 115 litres

Volume of on-board software:

A310 = 4 Mbytes   A320 = 10 Mbytes   A340 = 20 Mbytes

Number of digital links:

A310 = 136   A320 = 253   A340 = 368

On-board calculation power:

A310 = 60 Mips   A320 = 150 Mips   A340 = 250 Mips

Conclusion   In order to evaluate the growth potential in environmental conditions it is necessary to make an analysis of materials evolution and of electromagnetic environment. The size of the threat by EMI has increased by about 100 dB since the 1950s due to technological changes.

The materials evolution will have to take into account the fact that composites will keep being used on aircraft as they will be better mastered and as their cost will be lower (thermoplastic). However, on the contrary, the breakthrough of aluminium-lithium alloys is not yet a certainty. This material evolution will influence environmental conditions (temperature EMI, RFI), and in the same way radio guidance, for example.

Generally speaking, as for equipment, the invasion of numerical technologies of the last 15 years will go on. It will imply a broadband transmission support working in a disturbed environment (diaphonics).

3.2   Task 2: review and assessment of system architectures, components and procedures

Introduction   From the system requirements identified in Task 1, a hypothetical avionic system and the associated weightened requirements have been defined in Subtask 2.4. On this basis the candidate architectures using the state of the art (SOA) of the optical technology identified in Subtasks 2.1, 2.2, 2.3 and 2.5 have been assessed and scored, leading to a first selection of the most suitable architectures for aircraft applications. In parallel, the fibre optic components handling procedures for manufacturing, installing, testing and repair of harness have been more deeply reviewed and assessed in Subtask 2.5 and will be demonstrated in Subtask 3.2.

Four of the subtasks are focused on the review of the different subjects to be investigated to install an optical system in aircraft. All of these subjects are firstly assessed individually and secondly combined for the general assessment of the possibilities.

The subjects to be investigated were defined as follows.

Subtask 2.1 Review of data system topologies. Responsible partners: MBB, ALENIA, FARRAN, AEROSPATIALE, SMITHS, STC, CERT, NTUA, FOKKER, NLR.

Subtask 2.2 Review and assessment of component modules. Responsible partners: MBB, ALENIA, FARRAN, SMITHS, STC.

Subtask 2.3 Comparison of protocols for databus systems. Responsible partners: MBB, ALENIA, AEROSPATIALE, SMITHS, STC, CERT, FOKKER, NLR.

Subtask 2.4 Assessment and selection of system architectures. Responsible partners: MBB, ALENIA, AEROSPATIALE, SMITHS, STC, CERT, FOKKER, NLR.

Subtask 2.5 Review of fibre handling procedures. Responsible partners: MBB, AEROSPATIALE, BAe, AMD, STC, FOKKER. After the definition of a hypothetical aircraft system by the use of the outputs of Task 1, the above subjects are merged for the assessment of the possible architectures.

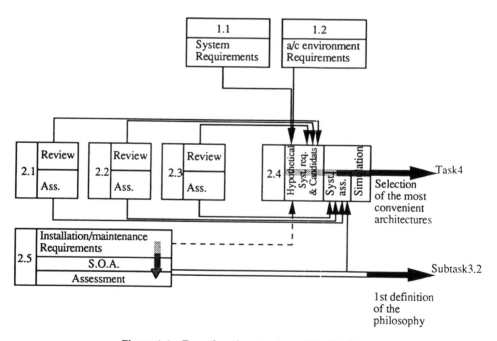

Figure 1.1   Describes the structure of the Task 2.

3.2.1   Subtask 2.1: review of system topologies

Activities   The goal of the Subtask 2.1 is to review both active and passive topologies and to compare roughly their advantages and disadvantages. In order to be consistent and because of the early start of this task in the programme, the assessment is based on a common simple hypothesis including the performances, the cost and the reliability of the single components and the most important requirements of the aircraft avionics. The comparison criteria are the following: maximum number of terminals that can be supported, number of components employed (both passive and active), propagation delay, reliability of the network, installation and maintenance of the system, cost of the system, possible access methods that can apply, potential growth of the number of terminals supported (if the future trends call for it), and testability of the system.

Results   One of the more important requirements appears to be the number of terminals that the network can support. This constraint eliminated the passive ring topology which can support only seven subscribers. The linear T coupled buses are able to interconnect only a restricted number of terminals (around 20 subscribers) due to the very stringent power budget limitations, and further improvement of components is required before the use on a/c. For the same reason the expansion ability of the topology is limited. However, it has not been totally eliminated because the need for 50 subscribers depends on the avionic philosophy.

Apart from this, the reliability is one of the most important features that the application calls for; the system must be extremely reliable, so that its operation in the case of component failure during flight is ensured. This parameter has been assessed by the calculation of the MTTF of a link between two subscribers and by the calculation of the availability ratio which represent the percentage of usable subscribers after a single, double or triple failure on the network. This second parameter was introduced in order to take into account the redundancies which are always needed in avionics.

The linear T coupled buses (see Appendix 5.1) are able to interconnect only a restricted number of terminals due to the very stringent power budget limitations. For the same reason the expansion ability of the topology is limited. The reliability of the topology is not low, as the failure of any element of such a system does not disable all the terminals; the availability of the system is in the worst case of a single failure 50%. This is attributed to the fact that no active elements are used along the databus; thus the worst case is a break of the bus fibre cable that is common for all the subscribers.

The active ring (see Appendix 5.2) is relieved from power budget limitations because of the use of repeaters at each node. The problem with it is the high vulnerability to node or fibre cable features. The reliability of the active ring can be enhanced by some redundancy scheme of the media or the active components; such methods have already been used with success (e.g. FDDI). An option of the ring topology which aims at the increase of the reliability is the active ring with passive redundancy. This option can tolerate the failure of one node only without an unacceptable increase of the BER (this has, of course, to do with the performance characteristics of the emitter and receiver) but it seems difficult to identify a convenient protocol for such a topology. In general the ring topology offers convenient cabling, ease of installation and mainten-ance, flexibility and expandability; also simplifies the access to the media in combination with good network performance parameters.

Complete mesh topology uses a pair of active components for each possible communication. This approach totally decentralizes the media and the compo-nents, but leads to very high cost, installation and maintenance problems, and expansion inability.

The distributed star topology (see Section 3.4.2) presents very interesting features. It can accommodate a high enough number of terminals. Its reliability and availability are increased because of the distribution of the terminals to more than one component; as a result, the failure of a component does not put the whole system out of order. Also, the cost of this topology is reasonable.

Passive stars (see Appendix 5.3) face power budget limitations, but not to the same extent as linear systems. The power budget limits the number of terminals to a certain level, thus allowing no expansion of the system. The reliability of a passive star network is not as high as that of a linear system due to the use of the passive star coupler; a failure of the passive star coupler causes major problems to the whole system. A passive star coupler affects also the ease of installation and maintenance. Installation and maintenance difficul-ties arise from the centralized cabling as well.

Active star systems can support a large enough number of terminals, but the use of active elements in a centralized role puts the reliability of the system in danger. Installation and maintenance of such systems is hard; this is associated with the active components and their importance, as well as to the excessive cabling that connects the terminals to the central active node. The performance of these systems is rather poor, as the centralization of the active node and the broadcasting function it carries out leads to long delays.

3.2.2   Subtask 2.2: review and assessment of component modules

Activities   The Subtask 2.2 deals with optical components and modules. In order to design and implement optical data transmission systems it is neces-sary to understand the specific capabilities and characteristics of the various component technologies that will be used in those systems. To give the most realistic transmission system design, it is important to take as a starting point the currently attainable component capabilities, and that is what a large part of this subtask is concerned with. However, as the networks being considered are not likely to find their way into production aircraft for a number of years it is probable that component development will extend the capability of existing component technologies and may even produce new ones. It is therefore realistic to extend the assessment to try to predict what these future developments may provide by way of increased performance and capability, and this is also covered in the report.

Results The report distinguishes between passive and active components and studies each type of component in turn. One of the most important criteria for components is the ability to maximize the optical throughput of the system for the complete avionic environment and so to maximize the number of terminals supported by the network. This will be achieved by either maximizing the power launched onto the network, minimizing the loss on the network or maximizing sensitivity at the detector. Most components available at present are derived from commercial or industrial (other than aeronautic) designs, and their performance suffers when exposed to the more demanding avionic environment. However, manufacturers are beginning to develop the necessary components as the installation of fibre optics on producton aircraft comes closer.

3.2.3 Subtask 2.3: comparison of protocols of data bus systems

Activities In Subtask 2.3 different standard medium access control (MAC) protocols are compared with respect to their suitability for future avionic applications. The assessment criteria are the general performance (i.e. control mode, message latency, priority, throughput, fairness, broadcast and multicast addressing, classes of service, fault recovery mechanism) and the suitability to the optical medium. Each of the MAC standards analysed incorporates specification for the physical layer (PHY) appropriate for the MAC protocol. The PHY layer specifies the signalling characteristics, the physical interface to the transmission medium and the network topology. Also, if many of the PHY layers specify different electrical media, future fibre optical applicability is investigated. The IEEE802.3 standard protocol is based on the carrier sense multiple access with collision detection (CSMA/CD) mechanism. IEEE802.3 employs a fully distributed control and provide broadcast and multicast address modes; it is modular, flexible, and accommodates system growth smoothly as the system evolves. The throughput obtainable with IEEE802.3 is acceptable (up to 90% of the transmission rate), but it falls off when the traffic load or the number of subscribers increases (see the simulation results of Subtask 2.4).

Results CSMA bus IEEE802.3 has no provision for acknowledged message transfer, message priorities, or periodic/aperiodic message handling. However the major disadvantage of the IEEE802.3 is that it does not provide a deterministic upper bounded message latency. The PHY layer specification incorporated into the IEEE802.3 standard provides different electrical transmission media, however no limitation arises as a result of using optical transmission. Although optical collision detection mechanisms are difficult to implement (especially with high operational dynamic range), an optical standard IEEE802.3 is presently under development.

The IEEE802.5 (token ring) and the FDDI standard protocols are based on coordinate token capture mechanisms; they employ fully distributed control, provide eight (nine for FDDI) different message priority classes (hence the possibility to handle periodic/aperiodic messages), a message transfer acknowledgement service and broadcast and multicast address modes. Belonging to the deterministic protocol class, the IEEE802.5 and the FDDI provide fair access to the network and upper bounded message latency. Moreover the throughput is stable and increases with increase in network load. The FDDI standard provides a hard fault recovery mechanism, while in IEEE802.5 it is incorporated in the upper layers. The two standard protocols differ mainly in the PHY layer specification.

The IEEE802.5 PHY layer specifies an electrical transmission medium but an optical version of the IEEE802.3 standard is under development. The FDDI PHY layer specifies an optical dual counter-rotating ring operating at 100 Mbps. Although practical implementations already exist, this standard has not been finalized.

The IEEE802.4 and the AS-4074.1 token bus standard protocols use a coordinate capture token mechanism which is implemented on the logical ring established between the stations connected to the bus. They have similar performances, employ fully distributed control, provide four different message priority classes, the message transfer acknowledgement service (only IEEE802.4), broadcast and multicast address modes, fair network access, and upper bounded message latency. They also provide a command/response service, and ring maintenance operations which include: initialization, station insertion and removal, network dynamic reconfiguration after fault. The optical versions for both the IEEE802.5 and AS-4074.1 standards are under development. The Arinc 629 standard protocol is based on a carrier sense multiple access collision avoidance (CSMA/CA) mechanism. Arinc 629 employs a fully distributed control and provides broadcast and multicast addressing, three level of message priority, periodic and aperiodic messages handling, fair access to the network and deterministic bounded message latency (which is a function of the message length). Arinc 629 is being studied by several aircraft manufacturers (mainly Boeing) and is expected to be used as a future data bus for avionics. It specifies an electrical transmission medium and a transmission rate of 2 Mbps. Its compatibility with the use of optical fibre media for future higher speed applications is under consideration.

The last standard protocol analysed in Subtask 2.3 is the prEN 3910. It uses the time division command response multiplex mechanism. Presently used in EFA military aircraft, it specifies different physical media: hybrid (a high speed optical data network and a low speed command network in a separate wire STANAG 3838 databus), dual optical bus or two channel wavelengths multiplexed in a single optical bus. prEN 3910 employs centralized control and provides broadcast and multicast addressing, deterministically bounded message latency, and fair access to the network. prEN 3910 has no provision for acknowledged message transfer, message priorities (the bus controller software adapts the sequence of the messages according to the situation of the system). However the major disadvantages of the prEN 3910 are that it can directly address up to 31 stations and that its low speed command network works at 1 Mbps.

Many of the standard protocols analysed in Subtask 2.3 have the characteristics needed for consideration as valid medium access protocols for the first ODT system (Subtask 2.4); the only exceptions are the IEEE802.3 and the prEN 3910. The main reasons for discarding IEEE802.3 (CSMA/CD) are its unpredictable message latency and its unfair access to medium, while prEN 3910 should be discarded for its low number of subscribers and its centralized control (and hence vulnerability of the system).

Finally, it should be clear that none of the MAC standards reviewed in Subtask 2.3 satisfy completely the next avionic systems requirements. They only provide the lowest layer of functionality as defined by the ISO-OSI reference model. The missing communication functions, such as fault tolerance options, dynamic reconfiguration, congestion control, data security and routing between subnetworks, must therefore be allocated to the upper ISO-OSI layers.

The fourth subject to be treated is more an aircraft (a/c) manufacturer problem than a technical one and this is the reason why they are mainly involved in this subtask (11/11.5 mm). Because of the very limited experience available of fibre optic handling, all the procedures for the implementation of an optical harness have to be defined. These procedures should take into account both a/c manufacturer and a/c company constraints including tools, personnel skill, security and cost.

For the first time, it was planned to include a practical demonstration of optical harness manufacturing, installation and repair, but because of tight schedule and workload constraint, this demonstration has been postponed until Subtask 3.2.

This review starts from the electrical state of the art and analyses the differences between electrical and optical technologies for the production steps. The main identified differences are the level of protection of the optical end face against all types of pollution or damage, the importance of respecting the optical cable bend radius during installation and, more generally, the higher susceptibility of the optical harness to all types of damage that could appear during the implementation process (including transportation).

Concerning the components, it appears that environmental constraints are difficult to recover but this subject concerns Subtasks 2.2 and 3.3. From the implementation point of view the packaging of vulnerable components like couplers must be made rugged and some interesting components like splices or switches do not appear to be mature enough for onboard use. However, cables and connectors (mounted with the pot and polish technique) are near to the a/c requirements.

As expected, the tools have to be developed (or invented) for aeronautic use because optical technology is presently used only for ground application. They are not adapted to restricted space and they do not allow one to test a complex network by the use of a small number of input ports. For the implementation of the optical harness, three philosophies have been analysed: manufacturing in the workshop (i.e. loom board) and proceeding to install the complete harness on the a/c; buying the harness complete (off-the-shelf) and installing the harness on the a/c; and manufacturing the harness in situ (on the a/c).

The first two philosophies seem to be the most suitable from the points of view of time consumption, manufacturing quality, tooling performance (space restriction), level of skill required and the testing facilities. Installing/manufacturing the harness in situ is not recommended because of the perceived difficulties of achieving reliable terminations due to the greater risk of contamination to the fibre end faces, the high skill levels needed and difficulties imposed by the space limitation.

The final choice will depend on the mock-up installation that BAe and FOKKER will realize in Subtask 3.2. The preferred philosophy is likely to be a combination of the above proposal by the use, for example, of in situ cable termination when the routing of the terminated harness through a very restricted area is too difficult.

3.2.4   Subtask 2.4: assessment and selection of system architecture

Activities   Subtask 2.4 is the key point of the project from the system point of view and it should allow the partners to identify the most likely architecture(s) for the future. This work is split into two major parts: the assessment and selection of an optical data transmission system architecture, and the simulation of selected system architectures. During the working development of the first part, several problems appear (technical and methodological) that lead us to think that the results have to be considered very carefully. The goal of the Subtask in the project ODT Aero 0012 was the selection of system architectures, but, in fact, the results of the first part of the work are: a procedure that helps the system designer to find the preferable architectures, and an identification of the principle advantages and disadvantages of the studied architectures.

From the outputs of Task 1 two hypothetical systems have been defined. The first system (A) includes the most critical functions leading to a medium number of subscribers (35/45) and a high reliability level, and the second system (B) includes the other subscribers leading to a high number of subscribers (70/80) and a lower reliability level. These systems take into account existing avionics characteristics and extrapolated values for the future. However, it is very difficult to define these values because they depend strongly on the chosen philosophy. For example, the use of smart actuators hardly increases the number of subscribers and the data rate; the integrated modular avionics concept lead to a small number of subscribers (cabinet) and to two very different buses (the backplane with high data rate and very short distances and inter-cabinet bus with medium data rate which cover the whole a/c). The influence of the introduction of new systems like TCAS or ELS is not very well known. Therefore, the introduction of a hypothetical system inevitably reduces the representativeness of the system.

Another point which limits the architecture selection process is the fact that such a process uses a loop back with the subscriber characteristics (functions included, security level, needed redundancies, . . .), which are not defined for a hypothetical system!

In this context the two hypothetical systems are based on the philosophy of the A320 avionics, the only well known flight-by-wire a/c with increased data rate, number of subscribers and reliability, and, as expected in the future, dispatch of the subscribers within the whole a/c.

Results An ideal method for the assessment of such a complicated thing as avionics does not exist. The method used for Subtask 2.4 tries to take into account the risk aspects while taking advantage of scoring aspects. After the precise definition of the requirements, it is decided whether they are applicable to topology, protocol or components independently or combined. Next, requirements are classified into exclusion requirements (of pass/fail nature like the mandatory number of subscribers), selection requirement (like cost or weight of the harness) or both for selection requirements which have to reach a minimum value. All the selection requirements are weighted for the assessment process.

Only the passive ring and the reflective star topologies have been discarded by exclusion requirement because of the very low number of supported subscribers. A linear bus topology that does not reach the minimum number of subscribers (25 instead of 35) is not discarded because of the expected improvements and because of the redundancy needs (which cannot be assessed due to the missing loop back) that could lead to a reduction of the number of subscribers connected to the same physical bus.

The main difficulties appear with the reliability requirements. The first technical problem is the weak data base of components MTTF. Some values exist for telecommunications applications but it is nearly impossible to find information about components under aeronautic environmental constraints. The second problem is about the method and the ODT partners do not agree. The applied method is firstly to consider the elementary system without any redundancy and to calculate the level of redundancy needed of this elementary system on the basis of architecture only (without taking into account the reliability of the subscribers). The second (unapplied) method is to put hypothetical values for triple or double redundant subscribers on the basis of A320 systems, and to calculate whether the requirements are met and to assess the needed modification on basic topologies to reach these requirements.

On this basis, the architectures which obtain the best scores: for the system A are star topologies (1st centralized, 2nd distributed) with a token bus protocol and without special components (the most complicated component is a 50 × 50 ports star coupler); for the system B the best scores are obtained with active ring topologies (1st FDDI, 2nd single ring), token ring protocol and one critical component for the FDDI: the active electro-optic switch.

Because of the above problems and of the duration of the simulation, the data bus configuration has been chosen before the completion of the architecture assessment. The simulated topology is the central star because it is the most constraining passive topology. The protocol models developed during the ODT programme are the token bus and the CSMA/CD. Because of time scale and work load, the token ring and the ARINC 629 are not simulated (it was not planned to develop all the possible protocol models within ODT!).

Two protocol simulation experiments have been performed (see Appendices 5.4 and 5.5).

In the first experiment, an ODT system under a load with a constant ratio of periodic and aperiodic traffic was studied for both token bus and CSMA/CD. The number of stations and the transmission rate were kept constant (32 stations, 10 Mbits/s, respectively). The bus load was varied by enabling or disabling stations to generate traffic onto the bus. Important conclusions are that for bus loads lower than 50%, CSMA/CD provides a lower average message latency than the token bus but a higher maximum message latency. For bus loads higher than 50%, both average and maximum message latency for the token bus are lower than for CSMA/CD. Furthermore, the maximum bus capacity used for message transmission is between 91% and 96% for the token bus and only 86% for CSMA/CD.

In the second simulation experiment, the performance of ODT systems was tested for two hypothetical avionics systems. Only the token bus media access protocol was investigated for this experiment. The first system is characterized by a low number of stations (32), and a large amount of traffic between the stations (4.75 Mbit/s). The second system is characterized by a higher number of stations (64) and a lower amount of traffic between the stations (1.24 Mbit/s). The traffic characteristics were defined to resemble actual traffic characteristics as much as possible. The values for TRT and THT were selected according the rules that gave the best performances in the first experiment (TRT = period of message stream with shortest intermessage time, THT = TRT/number of active stations). The bus load was varied by changing the transmission rate. The simulation results show that the average message latency remains acceptable (smaller than 15 ms) in the first system when the bus load is kept lower than 60%. The second system displays acceptable average message latencies (around 10 ms) when the bus load is kept lower than 50%. For further work it would be interesting to develop other protocol models like ARINC 629, token ring (IEEE 802.3, SAE-AS4074.2, FDDI), or token bus (SAE-AS4074.1), to incorporate mechanisms which introduce faults such as bit errors and link failures and to extend the models with fault recovery functions. This would allow one to validate fault recovery mechanisms of ODT systems and help the optimization of system parameters. Indeed this task does not allow one to select the best architecture; it give information to the system designer for the selection of the preferred architecture for a real system. The rest of this work is in Subtask 3.1: system design guidelines.

3.2.5 Subtask 2.5: review of fibre handling procedures

Responsible partners: MBB, AEROSPATIALE, BAe, AMD, STC, FOKKER.

Activities Subtask 2.5 describes an overall review of the fibre optic harness handling procedures. This document first analyses the different philosophies of fibre optic harness manufacturing. These philosophies are:

(a) manufacturing in the workshop (i.e. loom boards) and proceeding to install the completed harness on the aircraft

(b) buying the harness complete (off-the-shelf) and installing the harness on the aircraft

(c) manufacturing the harness in situ (on the aircraft).

The most suitable philosophy to be used for installation is likely to be either (*a*) or (*b*), from the point of view of: time consumption, manufacturing quality (optical attenuation, length of cable, . . .), tooling performance (space restrictions) and level of skill required, and testing facility.

Concerning the harness installation on aircraft, the main identified differences between fibre optic harness and electrical harness installation are: the level of optical protection against all types of pollution, the importance of respecting the optical cable bend radius during installation, and the necessity (as much as possible) of installing optical harness after electrical harnesses to avoid optical damage during installation.

Results Concerning the component's packaging and the required tooling, it appears from Subtask 2.5 that: vulnerable components of harnesses (e.g. couplers) must be made rugged and if possible must be enclosed within a box with connected outputs, existing tools made for laboratory use will have to be made rugged and hand-held for aeronautical use, and a new family of tools has to be developed.

One problem area may be the transportation of the harness from one location to another before it is installed in the aircraft. Concerning the assessment of techniques, the most advanced techniques for terminating harnesses is the 'pot and polish' technique. Nevertheless, dedicated aeronautical tools still have to be developed, and research concerning epoxy's sensitivity to ageing, humidity and temperature needs to be continued.

The major concern for installation of pre-built harnesses is the risk of damage during transit, during installation of the harness or during subsequent aircraft building operations.

Installing/manufacturing harnesses in situ is not recommended because of the perceived difficulties of achieving reliable terminations due to the greater risk of contamination to the fibre end faces, the high skill levels needed and difficulties imposed by the space limitations imposed by the installation.

Conclusion   The final choice depends on several factors that may indicate different priorities, considering the circumstances of the installation. For example, installation on a wing may require a mixture of all philosophies: a harness may be bought or built in the workshop; space restrictions at the wing root may require termination of that end of the cable off the aircraft; at the wing tip it may be necessary to terminate the cable in situ, because of difficulties in routing a terminated cable through the wing due to space restrictions (size of connector(s)).

To determine which installation philosophy is preferred on the basis of the theoretical study carried out in the previous sections of Task 2.5, the three philosophies must be compared for each aspect and rated in order of merit.

For detailed information the reader is referred to the final report Task 2 subtask 2.5, e.g. Table 2.5.6-A shows that on the basis of a total of the ratings, and consideration of the weighting (similar to that used in Subtask 2.4) buying a pre-built harness is considered the most desirable approach for the airframe manufacturer.

3.3   Task 3: establish implementation guidelines for bus systems

Task Group Leader: Dassault Aviation
Responsible partners: Alenia, AS, Fokker, MBB, Dassault, AS, BAe, Farran

The work provided in other tasks and subtasks of this ODT project was intended to define a procedure to select the relevant architecture for a specific optical data transmission system. The various parameters of this system, especially the number of subscribers, their locations in the aircraft, the type of aircraft and the data transmission characteristics, were fixed *a priori* in order to limit the field of investigation for topology, protocol and component selection.

The purpose of Task 3, entitled 'bus system implementation guidelines' is to give to any system designer some guidelines for the design of the ODT system, for its installation on aircraft and for the selection of optical component matching with the system specifications.

3.3.1 Subtask 3.1: system design guidelines

Responsible partners: Alenia, MBB, Dassault, AS, BAe, Farran

This subtask intends to define a general procedure for the design of an optical data transmission system. It should enable a system designer to converge to relevant architectures according to identified parameters. It was not conceivable to define, with the allocated time and manpower, a numerical selection procedure (as done in Subtask 2.4) for all the possible parameter values and for all topologies, protocols and components. Moreover, such a selection procedure could not have been objective. Thus, the purpose of the involved partners has been to give guidelines for each step of the system architecture definition.

Activities The performance of the optical data transmission system is linked to the selection of topology, protocol and components. When designing an ODT system the aircraft manufacturer should consider the following aspects and compare them to his system requirements.

The following topologies have been assessed: linear T coupled bus, ring (active and passive), and mesh and star (active and passive), taking into acount comparative calculations of optical power budgets according to the number of subscribers in the system, their statistical repartition in the aircraft, component optical performances and a desired growth potential.

Now consider protocol analysis (the format of the message transmitted on the data bus and the service mechanism offered to the subscribers). The protocols IEEE 802.3, IEEE 802.4, IEEE 802.5, FDDI, AS-4704.1, MIL-STD 1553 and ARINC 629 were analysed. This analysis should define word length, message length, address modes, class of services, priority mechanism, message latency, data rate, error detection mechanism, fault tolerance, bus overload tolerance and facilities for tests and maintenance.

The compatibility of available optical components with the topology and protocol parameters and environmental conditions for the various areas of the aircraft was assessed.

The relevant reliability parameters of the ODT system are the undetected bit error rate, the detected bit error rate, the mean time to failure, the mean time between failure, the recovery time and the level of redundancy of the system. The reliability analysis defined in this subtask for these various parameters proceeds through the following sequence.

Theoretical analysis of the hardware including prediction of component reliability, calculation of system reliability with or without redundancy, failure analysis for components or human intervention.

Selection of components according to reliability performances, qualification for airborne applications and analysis of particular failure mechanisms.

Maintenance analysis taking into account availability and ease of use of diagnosis and reparation tools and associated procedures which determine meantime to repair and average time for repair.

Quality control of software which includes reliability study through failure simulation tests and debugging programs.

The environmental conditions in the aircraft zones in which the ODT system will be implemented are critical for performance and reliability of this system. The temperature range is of great concern for optical applications. The sensitivity of the various optical components at the current state of development justifies the use of tight buffer cables, light emitting diodes in the first transmission window and PIN receivers. Little is known about the other component behaviour with temperature variations.

Other limiting environmental conditions for components are vibration, humidity (which requires equipment shielding), contamination (especially for optical contacts within the connectors), salt spray (which requires protection of electronic modules) and ambient light (which can be picked-up in the fibre). The component manufacturer has to guarantee the quality of the components under all the specified environmental conditions.

The installation and maintenance guidelines given in this subtask only focus on the optical specificity. Their purpose is to be applicable in and outside pressurized areas, to be of high reliability for all environmental conditions, to ensure 100% dispatchability during aircraft life and to provide easy maintenance, repair or test procedures. These guidelines concern:

- routing (number of connections, bend radius, segregation, location of sensitive components in protected areas, cable extra length storage and accessibility)

- marking (readable, resistant and non-damaging identifiers of cables, connectors and input and output ports of couplers)

- testability of the harness (access points, compatibility of connectors, potential use of OTDR and built-in-tests)

- cleaning (accessibility of the harness, adapted tools, physical protection for the handworker, relevant inspection tools and preferential use of male–male connectors)

- tooling (optical component limitations and aircraft structural limitations)

- handling precautions (transportation, routing, clamping the harness and protecting the end tips of connectors against pollution)

- safety requirements for the personnel (mechanical and optical risks)

- time saving guidelines (use of standardized components, automated tools, spare pieceparts for quick repair and built-in test for maintenance)

The costs for the implementation of an optical data transmission system in an aircraft integrate the following.

Harness manufacturing costs including the cost of optical components (which should be minimized by reducing the number of suppliers and using standardized products and tools), the cost of production facilities (including production equipments and power supply), the cost of tooling associated with optical handling and testing, the cost of personnel training to the optical harness specificities and the cost due to harness average manufacturing time.

Installation, maintenance and repair costs concerning personnel training (precautions to be taken in comparison with electrical harnesses), tools that should be adapted to the optical characteristics (bend radius sensitivity) and time for installation.

Costs of the ODT system implementation including the reliability of the system, which is mainly determined by the lifetime of the active components and which gives the mean time between repair and the average time for repair.

Other costs associated with the storage of spare pieceparts, the transportation of the complete harness from the workshop to the aircraft installation implant and the safety measures which require special equipment (protective gloves and glasses).

3.3.2 Subtask 3.2: manufacturing and installation guidelines

Responsible partners: AS, BAe, Fokker, MBB, Alenia, Dassault

In Subtask 2.5 of this ODT project, entitled 'review of fibre optic handling procedures', two main philosophies of manufacturing and installing an optical harness in an aircraft have been identified, which are manufacturing the harness in a workshop or manufacturing it in situ.

In order to adapt the manpower allocations to the work on this subtask, the involved partners decided to focus on the solution which appeared to be costless and more adapted to industrial production, that is to say manufacturing the harness in a workshop and installing it afterwards on the aircraft.

As mentioned in the Task 2 activity report, the practical demonstrations planned in Subtask 2.5 have been reported to Subtask 3.2 with a re-allocation of manpower. The objective of the subtask, as redefined in the progress of the project, is to give manufacturing and installation guidelines resulting from the analysis of the components and associated tools used for practical demonstration on mock-ups (Fokker 050 front fuselage and BAC 111 aircraft wing).

This subtask was then divided into three parts: proposed guidelines, practical demonstrations, and analysis of the practical demonstrations.

Activities The first part describes harness implementation proposed guidelines. For each type of component, these recommendations try to cover three

main aspects: the quality which should allow to preserve the optical performances, before, during and after assembly and during the installation; the
security for the personnel, in order to avoid accidents due to the tools, fibre
fragments and optical radiations; and the maintenance, repair and replacement
constraints such as, failure detection, required skill level or accessibility.

The second part of the work consists of the practical demonstrations. The
first one took place on a Fokker 050 front fuselage mock-up and the second one
on a BAC 111 wing mock-up. They intended: to verify the proposed guidelines
by comparing two different approaches; manufacturing the harness in workshop before installation or installing the optical cables one by one directly on
the mock-up; and to assess the different aspects of:

- manufacture of the optical harness,

- its installation on aircraft,

- repair,

- required skill,

- tools,

- tests.

Fokker followed the first approach and manufactured their harnesses, including both electrical and optical cables, in a workshop before installation. BAe,
on the contrary, installed the optical cables, one by one on the mock-up. A lot
of them were terminated in situ. The goal of these demonstrations was to
assess the ability of the optical components to stand the 'cable is cable'
philsophy (i.e. the implementation of optical cables in aircraft like electrical
ones). However, cable overlength to allow contact replacement, restrictions
concerning supporting cables at the strain relief of connectors, in order to avoid
fibre kinking or folding, also restrictions for bundling at crosses and branches
and more care for handling were advised.

The third part concludes with the analysis of the tools, techniques and
components used in these demonstrations.

Results These two demonstrations were very instructive and highlighted
several points. The manufacture of harness in a workshop before installation
seems to be more easily applicable to an industrial process than the other
approach. The 'cable is cable' philosophy does not raise any problems. The
optical passive components easily stand installation and repair. In particular,
FO cable handling does not need as much care as previously thought. No
additional loss has been measured after the installation has been done by
non-specialized people. As far as classical materials and tools for installation
are concerned, they can also be the same as those used for electrical cables.

Conclusion   However, a few weak points remain. In situ termination is hard to implement since it requires appropriate tooling, time and comfort. Over-length of cable seems to be of no use for repair procedures which remain too long and unpractical to perform in a realistic industrial process. Finally, test methods and devices presently used are not satisfactory and need to be developed and improved to reach better performances and fulfil all A/C requirements. These subjects, which need further study, could be dealt with in the second phase of BRITE/EURAM programs.

3.3.3   Subtask 3.3: component selection guidelines

Responsible partners: Farran, AS, CERT, Alenia, BAe, Dassault, HAI, STC, SI

This Subtask 3.3, originally entitled 'component design guidelines' in the contract, was renamed during progress in the ODT project as 'component selection guidelines'. This modification was motivated by the fact that the whole ODT project is oriented, considering experience of the involved partners, to airborne application of optical data transmission and not to the conception and manufacturing of new optical components.

Activities   The purpose of this subtask is to give to the ODT system designer (aircraft manufacturer) some ideas about the operating principles of every optical component, some accurate information on the various technologies used to make them and the performances of commercially available optical components. This information is compared with the ODT requirements in order that the system designer should be able to choose the appropriate components for his ODT application.

Results   Suitable optical fibres for aerospace applications are mainly multi-mode fibres, whether step index or graded index, and the relevant cable structure is the tight buffer. The corresponding parameters for selection are: environmental behaviour (temperature, microbending, flammability, OH sensitivity), optical performance (bandwidth, attenuation, numerical aperture, core diameter), manufacturing characteristics (strippability of coating, type of material, length control), installation parameters (marking, tensile strength, bend radius limitation, protections) and maintenance limitations (accessibility of fibres in the cable, repairability, standardized products).

Two types of connectors are candidates for ODT applications. They are butt joint connectors with cylindrical or biconic sleeve and expanded beam connectors. Key parameters of connectors are insertion loss, return loss, susceptibility to ambient light coupling and crosstalk (for multiway connectors). Selection of connectors should take into account: optical constraints (performance, repeatability), assembly constraints (simple and rugged termination operation and connection process), installation constraints (component damaging avoidance, easy cleaning, mishandling control), environmental constraints (individual sealing, pistonning avoidance, no degradation), test and control constraints (mating/demating facilities, visual inspection), maintenance and repair constraints (mating durability, tooling), cost parameters (use of standardized components and tools) and lifetime of connectors (non-toxicity, non-corrosive, resistant).

Splices, like connectors, are used to interconnect optical cables or an optical cable with optical devices. The parameters to take into account when selecting a splice are: attenuation of transmitted power per splice (due to coupling, Fresnel refraction or misalignment), mechanical strength, environmental sensitivity, reliability and stability, repeatability of the splicing procedure, availability and ease of use of associated tooling and cost per splice. Splices can be divided into two groups: mechanical and fused splices. Considering the parameters mentioned above, mechanical splicing with crimping seems more adapted to aircraft applications, although it is not applicable in explosive areas. However splices should be used only for temporary repairs and replaced as soon as possible by a new cable, since there is still some uncertainty about their reliability within an aircraft environment over long periods of time. Improvements for future products should concern reliability, quick installation, ease of manipulation for in situ interventions, satisfactory protection against contamination of fibre end faces.

Tee-couplers are applicable to linear and ring topologies. Selection criteria concern tap on efficiency, tap off ratio, transit loss (or bus loss) and return loss (depending on the system characteristics). Selection criteria concern tap on efficiency, tap off ratio, transit loss (or bus loss) and return loss (depending on the system characteristics). The different technologies provide fused fibre couplers, mated fibre couplers, hard clad fibre couplers, miniature bulk optical tee-couplers or integrated optical tee-couplers. The selection procedure should consider the number of subscribers, the topology of the system, calculate minimum tap efficiency and optimum tap ratio and use figures to determine maximum allowable excess loss. According to this procedure, fibre-based devices currently seem to have the best optical performances.

Star couplers are used as the central power distribution in data bus network. A signal injected into any of the input ports is equally distributed to all the output ports. The selection criteria that should be applied by the system designer concern geometrical parameters (number of I/O ports, type of optical fibre cable, volume and weight, access to the system), optical characteristics (splitting loss, coupling ratio, excess loss, uniformity, insertion loss, directivity for transmissive couplers, backscatter, modal sensitivity, wavelength passband and achromaticity), environmental behaviour and cost. The different technologies available to manufacture optical passive star couplers either transmissive or reflective are: abrasion/glueing, mixer rod, fused biconical taper, integrated optics and micro-optics. Up to now, there has been little specific development of star couplers for avionic applications and the available devices still have to be implemented in protected areas of the aircraft (insertion loss sensitivity to temperature).

According to current performances of optical transmitters, the best suited components for ODT are LED emitting at 850 nm into multimode fibres. However, most of the available components still do not satisfy the temperature range requirements of ODT systems. Among candidate devices the following selection criteria are applied: optical performance (coupling efficiency, stability, insertion loss repeatability), installation characteristics (mounting electronical tuning reduction, fibre deterioration avoidance, easy cleaning), environmental behaviour (no long term additional loss with vibration, temperature and humidity), test and control ability (electro-optical control of transmitters), maintenance and repair facilities (mating durability, in situ intervention, easy cleaning), low cost (standardized devices) and lifetime performance (durability, compatibility with electrical output).

Receivers detect optical signals and convert them into electrical signals. The parameters to take into account for the selection of receivers are: surface of active area, wavelength, responsibility, quantum efficiency, dark current, capacitance, rise and fall times, operating temperature range and cost. The two main types of receivers are APD and PIN detectors. Although APD detectors offer higher performances (gain), PIN detectors seem more appropriate for present ODT applications where the distances are quite short because they are simpler, cheaper and less sensitive to temperature variations.

Repeaters are used to regenerate the attenuated and distorted signal. They integrate a receiver (photodetector, amplifiers), an electronic processing unit and a transmitter. The parameters used as selection criteria are: optical parameters (wavelength, sensitivity of receiver, dynamic range, power coupled into FO, bandwidth, propagation delay, BER), reliability parameters (MTTF, MTBF), environmental parameters (temperature operating range, vibration . . .) cost, size. The technologies used for their realization are progressing from discrete components to hybrid and monolithic integrated circuits. The performances of available repeaters in terms of bandwidth, sensitivity and environment seem to reach satisfactory level. However, progress can still be achieved as regards reliability, cost, compactness and weight.

Switches are used to reconfigure a network either for normal functioning or for a failure recovery mechanism. Considering available devices, electro-mechanical switches (moving beam or moving fibre devices) seem preferable to electro-optical switches (because they are mainly applicable to single mode fibres). The criteria applied for the selection of switches matching with the ODT requirements are: operational performances (insertion loss, back reflection and switching speed), environmental behaviour (temperature, shock and vibration) and reliability characteristics (actuation mechanism sensitivity, non-latching switches preferred to ensure fail-safe operation). Although existing devices prove to have satisfactory optical performances, environmental behaviour of switches for avionic applications still have to be improved.

3.4 Task 4: preliminary specification and selection of first generation of ODT systems

Responsible Partners: Alenia, AS, BAe, CERT, MBB, NTUA, Smith

This Task 4, originally entitled 'initiatives for component and module standardization' in the contract, was renamed during progress in the ODT project into 'preliminary specification and selection of first generation of ODT systems'. This modification was motivated by the fact that the whole ODT project is oriented, considering experience of the involved partners, to airborne application of optical data transmission and not to the conception and manufacturing of new optical components. Subtask 4.3 'initialization of standardization' was shifted to Task 7 'preparation of draft specifications'.

The subjects to be investigated were defined as follows.

- Introductory notes to: system requirements, selected topologies, power budget, and system specification.

- Elaboration of preliminary specifications for fibre optic components (cable, connector, coupler) and modules (transmitter, receiver) for selected topologies.

- The selection of first generation of ODT systems. It was suggested to carry through the selection of candidates for components/modules on the basis of: compliance to the specifications above, costs, easy maintenance, and reliability.

The task proceeds by producing the power budgets and system specifications for topologies which are selected to be compatible with the requirements of subsystem A, described in para 2.4.2.2.1. The power budgets and system specifications are produced for a first and second generation of FO (fibre optic) component technology.

In Subtask 4.1 the component and module specifications are then elaborated, which consist of two parts: generic specification, and a table containing the numerical values of specific system topologies and layouts for the first generation of an ODT system. The individual sections of this subtask are finished by giving recommendations for the second generation based on realistic assumptions for expected future developments.

The task cannot be considered as representing the solution for the optimum ODT system (which would require more manpower than presently planned) rather it shows the approach in establishing the component specs for an ODT system on the basis of selected topologies. The task further provides information on these properties of fibre optic components/modules which need further improvement in order to increase the potential of fibre optic systems relative to conventional electrical systems.

Activities   Two topologies have been selected to match the key system requirements: the central transmissive (passive) star, and the distributed transmissive star.

   For other versions of the distributed transmissive star topology the reader is referred to the final report task 2, Subtask 2.1.3.

   The selection of topologies for subsystem A (maximum 50 subscribers; 10 Mbit/s) has been carried out on the basis of the following main criteria:

(a) the number of terminals the topology can support within the available power margin emitter/receiver;

(b) mean time to failure (MTTF) of a fibre link between two terminals (taking the longest path for a ring, $N = 50$);

(c) mean time to failure of the coupling element which mainly determines the network availability.

The numerical values for MTTF as listed in the table have been taken from Subtask 2.1. The application of the criteria (*a*) and (*b*) restrict the number of applicable topologies to 4 or 5 with a preference to the central transmissive star and the active switch star configuration. The preference to the latter two topologies mainly results from the fact that the FO connector is the weakest component in a FO link and that the latter two topologies use less connectors in a terminal-to-terminal link than the others.

   The selection of topologies also has to consider the probability of a network

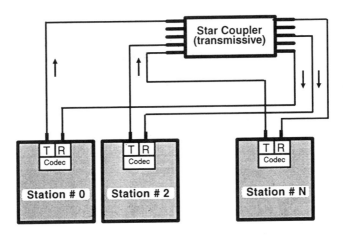

Figure 1.2   Central transmissive star topology

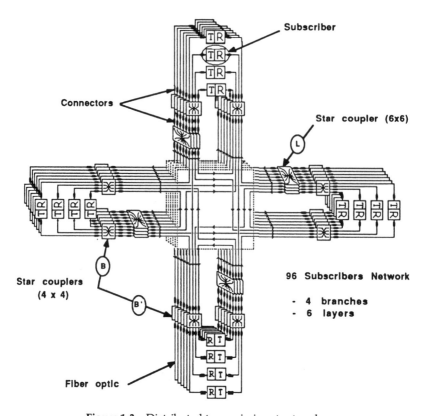

Figure 1.3 Distributed transmissive star topology

failure in the case that the important (central) coupling element would happen to have a failure. Therefore the third column of the table lists the MTTF of the coupling element. Here the topologies based on passive star couplers certainly have a high priority in comparison with the active star based topologies since the latter have a MTTF shorter by at least one order of magnitude (see also final report Task 2, Subtask 2.1). Correspondingly the active star based topologies are discarded when taking into account selection criteria C.

After having taken into account the criteria (a)–(c), the only topologies left over are: the central transmissive star, and the distributed transmissive star.

The central transmissive star-coupled network is simpler and has fewer FO components than the distributed star network; correspondingly the MTTF of a fibre link in the central star coupled network is twice that of the distributed star network.

On the other hand, in the case of a coupler failure the network availability is zero for the central star and still greater than 0.8 for the distributed star. This disadvantage of the central star network, however, has to be weighted with respect to the relevant MTTFs of the coupler and of the connectors. The MTTF of coupler and connector which are estimated to be 10^7 and 10^5 hours, respectively, are very different. Thus the appearing advantage of the distributed star topology is counterbalanced by failures of connectors which would occur much more frequently than a coupler failure.

For example at a time $t = 10^6$ h the probability of a failure of a central star network would be $t/MTTF = 0.1$; since at the same time already a substantial fraction of all fibre links have failed in a distributed star network due to connector failures, the expected number of communications possible becomes comparable for both types of network topologies.

Results for the power budget, see Appendix 5.6.

For the power budget of central transmissive star topology, see Appendix 5.7.

In order to have a first approach for a preliminary specification of components and modules, a power budget has to be made up (a transmissive star coupled network with up to 49 subscribers + 1 testport). The first column lists the losses which have to be taken into account based on presently available technology. The second column lists reduced loss numbers for the cable and the coupler; this loss reduction can be expected on the basis of ongoing development work.

Required power margin between transmitter and receiver	First generation (today)	Second generation
Cable losses (110 m) incl. temp. induced + bending induced losses transmissive star (50 port)	5.4 dB	3.5 dB
Max. insertion loss 6 connectors	22.0 dB	20.5 dB
Max. 6 × 1.2 dB	7.2 dB	7.2 dB
Repair/mismatch	1.0 dB	1.0 dB
Max. network loss (including all environmentally induced variations) Safety margin (including 3 dB for transmitter degradation over operational lifetime)	35.7 dB	32.2 dB
Required power margin between transmitter and receiver	41.7 dB	38.2 dB

In the following, the above numbers for the cable losses are derived. Again the first column lists the values derived on the basis of presently available technology.

Derivation of fibre optic cable loss (110 m)

Fibre optic cable loss	First generation	Second generation
Temperature-induced losses ($-55\,°C < T < +125\,°C$) (it is assumed that only half of the F.O. network experiences very low temperatures, ≤ 0.5 dB/10 m in the first generation	2.5 dB	1.1 dB
Bending losses (at R_{min}, max. 5 full turns)	0.8 dB	0.8 dB
Clamp losses (crush load losses)	0.5 dB	0.5 dB
Loss of straight cable ($110\ \text{m} \times 15\ ^{dB}/_{km}$, first generation, $110\ \text{m} \times 10\ ^{dB}/_{km}$, second generation)	1.65 dB	1.1 dB
Total fibre optic cable losses	5.45 dB	3.5 dB

A system specification is prepared which contains the main system parameters and describes the key optical parameters. Instead of listing the loss contributions of the individual components, the maximum and minimum optical path losses of a terminal to terminal link are specified. This procedure allows for a greater flexibility in varying details of a system topology; for example the reduction of insertion loss of a star coupler by using a smaller port number and taking the benefit of the available loss numbers by inserting an additional connector pair, without any need to modify the transmitter/receiver specification.

FO system, transmissive star
(50 subscribers)

System parameter

Data rate	10 Mbit/s
Encoding method	Manchester II
Signalling rate	20 MBaud
Number of subscribers (max.)	50
Message length	32 kbit
Frame period (min)	10 mins

Optical parameter

Fibre type	Stepindex, Multi Mode
Optical wavelength	770–870 nm
Transmission media attenuation (max.)	41.7 dB

Transmission media attenuation (min.)	18.5 dB
Output power (transmitter diode)	$+0 \, \mathrm{dBm} > P > +3.5 \, \mathrm{dBm}$
Sensitivity (receiver diode)	$< -41.7 \, \mathrm{dBm}$
Dynamic range (receiver)	$> 26.7 \, \mathrm{dB}$
Bit error rate (BER)	$< 10^{-10}$

Environmental conditions

Temperature range (operating)	$-55 \, ^\circ\mathrm{C} - +85 \, ^\circ\mathrm{C}$
Temperature range (storage)	$-55 \, ^\circ\mathrm{C} - +125 \, ^\circ\mathrm{C}$
MTTF (network)	$> 10^{+5} \, \mathrm{h}$
Ambient light coupling (at the receiver)	$-55 \, \mathrm{dBm}$ (max.)

FO system, transmissive star,
distributed network (50 subscriber)

System parameter (see before)

Optical parameter (see before except)

Transmission media attenuation (max.)	44 dB
Transmission media attenuation (min.)	19.5 dB
Sensitivity (receiver diode)	$< -44 \, \mathrm{dBm}$
Dynamic range (receiver)	$> 28 \, \mathrm{dB}$

Environment conditions: see above except

Ambient light coupling (at the receiver)	$-58 \, \mathrm{dBm}$ (max.)

General results and conclusions Data buses implemented in a ring topology (Appendix 5.2) have simple point-to-point links to connect subscribers, i.e. each subscriber receives messages from its predecessor and transmits to its successor. This topology allows high data rates in fibre optic as well as in copper technology, even if long distances have to be bridged between the individual subscribers. Each of the subscribers in a ring topology has to collaborate actively and correct, otherwise the communication will be interrupted. This problem was subject of numerous recommendations aiming to overcome the breakdown of individual subscribers. Active as well as passive bypasses have been proposed. An aircraft standard (using ring topology, e.g. HSRB) defines a second (dual redundant) counter-rotating ring for a higher degree of reliability. In this configuration each station is connected to two rings that support messages in opposite directions. Physically the ring topology does not form a real bus, rather the bus attributes are implemented at the protocol layer, e.g. by using token protocol. Standardization activities dedicated to aeronautics were seen in the SAE (Society of Automotive Engineers) High Speed Data Bus (HSRB AS4074) and in the AEEC (Airlines Electronic Engineering Committee) working in ARINC project 636 which defines FDDI (Fibre Optic Digital Data Interface ISO8802) for aircraft application.

  For the ringbus, the advantages are as follows: high data transmission rate,

large number of subscribers, suitable power budget, comparatively long distances between stations, low requirements for components, and easy repair.
The disadvantages are as follows: active system, total system break down
caused by the loss of one subscriber; preventive measure: bypass, counter-
rotating ring (expensive). High-speed repeater required by each subscriber.
Comparatively complex protocol.

In a linear fibre optic network implementation (Appendix 5.1), the separate
laying of the collection line for the transmitters and the distribution line for the
receivers, results in a very simple coupling technique. All bus protocols
commonly used in air and space vehicles were covered by the linear topology,
e.g. the protocols of the command response Milbus (MilStd 1553-B) and its
fibre optic equivalent (MilStd 1773), the European Standard EN3910, the Token
Bus HSDB (SAE AS4074.1) and the ARINC 629 modified for optical fibre.

The advantages are as follows: passive system, system remains working
even if one or more subscribers are lost; a great variety of protocols; low
expenditure for harness; and ease of repair. The disadvantages are as follows:
small number of subscribers; bad power budget; high signal dynamic at the
receiver; and high output power at the transmitter.

The physical nature of the star topology (Appendix 5.3) with respect to the
utilization of various protocols is fully identical to the linear topology; nevertheless it provides a better uniform power distribution, which results in a
suitable starting basis for a fibre optic implementation. Basically two different
topologies are formed by using a reflective star on the one hand and a
transmissive star on the other. The most suitable solution for aircraft is to build
the harness by using a reflective star coupler. In this case only one fibre optic
cable is connected to each subscriber. Harnesses built using transmissive star
couplers require two fibre optic cables to connect each subscriber; one cable
feed the transmitter and the other the receiver, which doubles the expenditure
on cables and connectors. If all subscribers are connected to one single star
coupler, this reduces the complexity of the harness. Star couplers equipped
with a low number of ports in a cascade lead to a more complex structure.
Nevertheless repair might be easier. An important design parameter is derived
from the loss budget. The lowest loss in the network appears. If a transmissive
coupler is used, then the signal power at the receiver basically arrives at the
output power of one transmitter divided by the number of subscribers plus
some harness losses. A much higher loss is observed at the harness built using
a reflective star coupler. In this case transmitter and receiver share one single
fibre, additional coupling losses of 6 dB arise from the required Y-coupler on
the subscriber side. Compared with networks using one single star coupler and
the same number of subscribers, harnesses built using cascaded star couplers
show higher losses. These losses arise due to: insertion loss of the couplers,
and additional losses due to a higher number of connectors or splices.

The advantages are as follows: passive system, system remain working, even
if one or more subscribers are lost; a great variety of protocols; big number of
subscribers (> 50); low signal dynamic at the receiver; and agreeable power

budget. The disadvantages are as follows: centralized star coupler, medium expenditure for harness, reduced length of network, high output power at the transmitter, and expensive to repair.

3.5 Task 5: development and definition of test methods

This task 5 was divided into the following file subtasks which correspond to each step identified previously for the test of components and complete and installed ODT, including built-in test. For each subtask the experience of the involved partners whether aircraft manufacturer, component manufacturer or research institutes, has provided substantial data on the various subjects.

Subtask 5.1 Identification of methods to measure, monitor and detect faults and deteriorations of performance. Responsible partners: CERT, MBB, AIT, AMD, FARRAN, SHMITH HAI, NTUA.

Subtask 5.2 Acceptance test methods of the fibre optic harness. Responsible partners: AMD, MBB, FARRAN, AS, STC, CERT, NTUA.

Subtask 5.3 Acceptance test methods for module interfaces. Responsible partners: SMITH, AMD, FARRAN.

Subtask 5.4 Definition of test methods for complete fibre optic system installed in aircraft. Responsible partners: MBB, AIT, AMD, AS, BAe, CERT, FOKKER.

Subtask 5.5 Definition of system self test requirements for performance deterioration and endurance monitoring. Responsible partners: Cert, MBB, Smith, NTUA.

3.5.1 Subtask 5.1

Identification of methods to measure, monitor and detect faults and deteriorations of performance.

Activities This subtask describes the method of measurement concerning the main parameters of active and passive components of a fibre optic network. All methods are well defined by telecommunication standards. As physical measurement, fibre geometry; or optical measurements, numerical aperture, attenuation, band width, and refractive index profile of fibre; as well for electro-optical measurement for receiver and transmitters.

Results For avionic applications no suitable complete European standards yet exist to cover qualification of components. Qualification involves testing the electrical and optical performance parameters of the device over the extremes of use conditions.

The environmental conditions in the aircraft zones in which the ODT system will be implemented are critical for performance and reliability of this system. The environmental conditions are those listed in Subtask 1.2. Reliability tests are not described in this subtask. They must be included in qualification testing. Reliability tests typically involve electrically, optically, thermally, and or physically stressing devices to determine long term stability, when operating under the specified operating conditions and extremes of the anticipated use environment.

3.5.2   Subtask 5.2: acceptance test methods of the fibre optic harness

Activities This subtask describes the test methods used to qualify and to accept the various passive optical components which will compose the fibre optic harness. The test methods for the fibre optic harness are limited to attenuation control, sufficient to ensure the correct manufacturing of the harness. The test methods for attenuation measurement, described in Subtask 5.1, are the insertion loss technique, and the backscattering technique. These acceptance tests are used to ensure that the initially established quality remains. The acceptance test will need a reference harness and a tolerance value for the attenuation of all harness branches.

Results Acceptance test methods of the fibre optic harness. A fibre optic harness is understood to consist of cables, connectors, couplers and splices as far as necessary to realize the complete FO network in the topology given by the ODT system specification. The tests must consider the individual components and the complete harness not yet installed. For tests of the installed system see Subtask 5.4.

Computer acceptance tests are fairly routine because they must also be carried out for electrical cables and connectors in aircraft; therefore the following paragraphs can make use of existing test standards modified for fibre optic techniques, for example IEC 793-1, IEC 794-1, IEC 874-1, BS 6558, ANSI/EIA-RS-455 or DIN/IEC 46E(CO)25, IEC 86(CO)136, 42, 48, IEC 68-1, IEC 540, prEN2591, CCITT G.651. However, of course, special test procedures arise from the optical character of the components, and as far as these are concerned many standards are still in draft status.

Acceptance tests in the narrow sense usually do not require the accuracy of qualification tests because they should only assure that a shipment of a product is in accordance with the specified properties of the component or module which had been demonstrated by the qualification tests at the end of the development phase. In a wider sense acceptance also comprises qualification approval at the end of development and has to be carried out only once, whereas production acceptance in the usual meaning is a routine repetitive task. Production verification tests are used for quality control and they require specialized equipment and knowledge.

Test procedures depend on the type of the equipment, reliability and cost aspects, confidence in technology, pre-qualification of material, etc. The extent of the test procedures, test lot sizes, test statistics, tolerances, accepted failure rates must be discussed again for every system.

Equipment and operator skill requirements for routine acceptance tests should be as low as possible to reduce costs at the aircraft or system manufacturers. In the cases where no compromises can be made acceptance tests can be carried out by the manufacturer under the supervision of the purchaser. The details of these procurement procedures are up to the individual companies.

Performance figures given for the test are to be understood as examples; they have to be specified explicitly for every system. The same applies to the citation of standards.

It is not always appropriate to carry out all conceivable tests; EIA-RS-455 FOTP, for example, describes several dozens of tests and procedures. It is the responsibility of the system designer to decide (usually in collaboration with the manufacturer) upon detailed test requirements.

3.5.3   Subtask 5.3: acceptance test methods for module interfaces

Activities   This subtask defines the acceptance and qualification tests, that would be necessary for optical transmitters and receivers and also details the test methods necessary to perform these tests. There are no currently available standards concerning the integral testing of optical data transmitter and receiver for airborne applications, indicating the test methods and procedures for verification of the parameters listed.

Therefore this section tries to identify some of the available standards, specifications or proposals for standardization. Qualification testing, reliability tests, lot-to-lot controls are usually performed by the manufacturer and checked by the supplier of the fibre optic systems before integration in aircraft.

The definition of test methods includes: the definition of the qualification procedure and associated means, the definition of the transmitter parameters to be measured, the conditions under which these parameters will be measured, the definition of test methods to measure these parameters, the list of existing standards for transmitter testing, and the definition of acceptance criteria.

Results The acceptance and qualification testing of the transmitter and receiver modules is a necessary part of the qualification of the avionic equipment. These tests are aimed to ensure that the devices are capable of operating to the design specification.

Existing electronic devices are qualified by carrying out standard procedures and test methods such as MIL-M-38510 and MIL-STD-883. These documents specify which tests must be carried out for particular application areas and then give the details of how to carry out those tests. However these standards do not detail the functional tests that are carried out as part of the screening process. The testing of electrical data transmission components is a straightforward procedure, and validation and test plans form an integral part of the standards. However the immaturity of optical data bus standards for avionic applications shows itself in a lack of test procedures for optical testing. prEN 3910 is probably the most mature of such standards and although in many cases it indicates what is to be tested there is very little detail of how to perform those tests.

Conclusion The qualification procedure will be performed by the component manufacturer with possible control by the aircraft manufacturer. This qualification of the optical transmitter will concern all the piece-parts of the component and in particular the light emitting diode (LED). All environmental conditions listed in Subtask 1.2 will be applied for qualification of the product, taking into account for the operating range the location in the aircraft selected by the system designer.

The acceptance procedure will be performed by the aircraft manufacturer after reception of the active component. It should involve a minimum number of testing devices and should not include all environmental tests for cost saving. These acceptance tests will concern the packaged complete transmitter (or transceiver if transmitter and receiver are integrated in the same box).

3.5.4   Subtask 5.4: definition of test methods for complete fibre optic system installed in aircraft

Activities This task will define the measurement parameters for completely installed fibre optic harness. Measurement techniques will be developed either from those determined in Subtask 5.2, or alternatively by a totally new approach.

The aircraft companies will define the methods and techniques which require adaptation to in-flight testing. In this task, a set of characteristic system parameters and the method for measuring these parameters will be defined. These measurements will determine the correct and reliable operation of optical data transmission systems under different environmental conditions.

In Subtask 5.1 parameters of fibre optic systems were treated with respect to their measurement in general, but with emphasis on laboratory testing. In this subtask will be discussed which parameters are to be considered for testing of the complete ODT system (harness and terminals) after installation, and if additional test methods are necessary. In order to derive the required procedures, possible system failures and their causes must be analysed. Different topologies may require variations in test methods. For example a ring topology allows easy and complete testing from one point; testing of topologies with active components like repeaters and active stars is far more laborious. Characteristics of field tests are worked out and finally a market survey of field usable test equipment is given.

Results The evolution of field tests tends towards a total integration in the system of the test procedures. This trend considers the higher reliability of built in tests due to the non-pollution of the system and in particular to the performance of the optical connectors. It also avoids the problem of the availability of testing devices and adaptation to ground environmental conditions, which contribute to increase the maintenance and repair costs, but the BITE (built-in test equipment) does not allow precise location or even in some cases identification of failure points and failure reasons.

Remark A general and not trivial problem of equipment for fibre optic applications is the interface of the instrument to the fibre optic system. Since direct coupling of transmitter and receiver diodes to the fibre under test is not practical outside the laboratory, the manufacturers provide pigtails with connectors at the instrument panel. These connectors are usually not the problem because as a rule the suitable version can be selected from the hardware options.

A greater problem is the type of fibre used for the pigtails because the manufacturers for marketing reasons, of course, take fibres in use for telecommunications or LANs/MANs, which may be a single mode or in many cases a gradient index fibre or a fibre with a diameter of 125 μm. In aircraft systems fibres of 100 and 200 μm diameter are currently preferred. Very often it is difficult for the manufacturers to adapt the pigtailing procedure to other fibres, or at least they charge extra costs; the equipment cannot be shipped from stock and delays may be encountered. A jumper cable generally does not solve the problem, because the internal fibre between connector and diode or coupler or whatever must also be adapted to the system fibre to avoid measurement errors or severe backreflections.

Conclusions All necessary parameters of the complete FO system can be measured after installation. Transmission failures and failures of the electronic modules can be detected comparatively easily by built-in test equipment.

Defects of the FO harness are more difficult to localize (depending partly on topology); either they can be attributed to a certain part of the harness by analysing the terminal status or they must be investigated by optical measurements.

This diagnosis of harness defects may be possible by OTDR in real time or by external measurement on the ground or by a less sophisticated power measurement.

For the assessment and selection of the sufficient and necessary set of test methods more experience with operating FO systems is necessary.

Test equipment is available on the market but improvements are possible for the hardware and necessary for the software.

It will be an important task for the next stage of ODT development to investigate the OTDR (optical time domain reflectometry) method (or otherwise) as test of complete OTD system.

3.5.5 Subtask 5.5: definition of system self test requirements for performance deterioration and endurance monitoring

Activities/results The advanced avionic architecture for the next generation aircraft will feature common signal and data processing modules. The common modules will incorporate VHSIC technology and feature extensive. Built-in test and error logging capabilities. The common module exhibits intelligence and communicates at a language level through system interface. LAN (local area network) chips with self test features are in development and are ready for application in future networks. For overall testing (harness might be included) a maintainance controller may be applied in future.

Conclusion Two basic approaches to collect test and maintainance data should be investigated in a next step of ODT development. Monitoring live data traffic on one hand, and small frames of artificial generated test traffic. An integrated OTDR (optical time domain refectometer) provides test parameters of the fibre optic harness during network operation.

3.6 Task 6: experimental activities: laboratory and flight demonstration

Subtask 6.1 Laboratory demonstrations. Responsible partners: MBB, Dassault, BAe, Smith.

Subtask 6.2 Flight test demonstrations. Responsible partners: Fokker, BAe, MBB.

Goal of the experimental activities The purpose of Task 6 is to check the feasibility of aircraft ODT with available components and present technology. The experimental activities have to investigate relevant technologies on several levels. The physical level of the activities is mainly represented by transmitters, harness and receivers. The preferred type of component in these activities is the aircraft type. On the protocol level the most relevant protocols are to be implemented and compared.

Feedback between Task 6 and the subtasks is intended. The laboratory demonstrations shall use procedures of relevant data link layer and medium access layer. These procedures are defined in the protocols of token bus and ARINC-629. The physical layer of the laboratory demonstrations partly uses components which are of aircraft type.

The flight demonstration activity is to be based on the installation guidelines (Task 3.2), the handling procedures (Task 2.5) and the measuring methods (Task 5.4).

3.6.1   Subtask 6.1: laboratory demonstrations (see Appendices 5.4 and 5.8)

Subtask leader: MBB. The laboratory activities are intended to demonstrate the feasibility of data traffic along a fibre- only network, meeting the performance requirements of an modern commercial aircraft and the specifications of the protocols token bus or ARINC-629.

The system will use available hardware, which is slightly modified as far as necessary. The necessary opto-electronics and fibre optic hardware, including transceivers, fibre/cable and couplers, will be integrated.

The properties of such a system are to be investigated, such as attenuation of the harness over various paths, performance of frontends and characteristics related to protocols and implementation of protocols.

The decision on architecture (Appendices 5.9 and 5.10) is based on require-ments of the ODT system as far as the components are available. The implementation of two different protocols should require nearly no differences in aspects of hardware. Timescale and budget did not allow close conformance to the outcome of other tasks, such as Task 4.2, the selection of first generation ODT systems.

System components were selected so that the implementation of two rele-vant protocols was possible using the same hardware.

The main characteristics of the architecture for the evaluation of properties related to the protocol are: reflective STAR with 16 ports, single fibre cable terminated with SMA- or HA-connectors, transceivers intended to ensure enough powerbudget in a 32-port, network, digital transmission using the encoding scheme of Manchester-II, peak data rate of 20 Mbits/second, data-link layer and medium-access control implemented on a VME-size board (a data-bus module controlled by programmable gate arrays and a microprocessor), and higher layers (as necessary) are serviced by the microprocessor or the workstation, which houses the VME-size boards.

Although the network allowed 16 stations, the actual activities had been carried out using three stations, due to the constraints in time and budget.

Separate setups and systems have to be used for the evaluation of path loss, dynamic range and bit-error rate.

The preparation of the laboratory demonstrations was carried out by performing the following tasks: selection and integration of the fibre optic network, adaption to analysis of quality of signal, adaption to bit error rate measurements, analysis of the functions of the specific protocol, design how to implement the functions of the specific protocol, iterative convergent redesign, re-encoding of firmware and software until meeting the requirements of the project ODT.

The operation of an ODT system running an adapted implementation of ARINC-629 was demonstrated during the PCC meeting in March 1991. The easily distinguishable and recognizable parts of the system were the fibre optic network, the stations containing the fibre optic frontends and hardware supporting the several layers of the communication system. The essential contribution to this laboratory work resided within the memories of the multipurpose databus modules. The functions of the protocol were executed by the firmware and software, controlling the programmable gatearrays and the microprocessor. The protocol activity was visible at the additional opto-electric-converto equipped oscilloscope (see Appendix 5.11).

The scope showed the typical waveforms of the optical signal. Separately, the setup for signal analysis and the setup for measurement of bit error rate were demonstrated.

Most of the measurements on this system had been carried out during the second quarter of 1991.

Following the experiments on the ARINC-629 protocol the multipurpose databus modules got another firmware, which executed the token bus protocol.

This system was operated during the PCC meeting. Additional to the display of the fibreoptic signal, the terminal showed messages indicating the names of the stations active on the fibre optic network.

The measurements on this system had been carried out during the fourth quarter of 1991.

Conclusions   The fibre optic network including transmitters and receivers proved to be satisfactory, though the powerbudget was tight. This tightness was caused by the low performance of some preproduction versions of components.

Analysis of the quality of the signal showed a quite satisfactory speed behaviour of transmitters and receivers.

The planned performance in terms of bit error rate was reached, though some of the components performed poorly.

The laboratory activities demonstrated the overall correctness of the concepts.

The topologies proved their usefulness, though the attenuation and the intertransmission dynamic range remain the main constraints.

A certain shortage of powerbudget of the communication systems was experienced, due to the fact that components performed mostly near the lower limits of their specifications. Furthermore, the delivery of transmitters and receivers was sometimes seriously delayed.

The potential of the protocols Token-Bus and ARINC-629 had been fully realized by usage of programmable single-board VME modules.

The activity of implementing and testing the protocols revealed some demands for extensions and/or modifications of the protocols.

The proposed bandwidth for an ODT (Task 1) of 10 Mbits/s was surpassed (20 Mbits/s) without seriously limiting the possible number of stations (up to 32).

Recommendations and further activities   Continuation of laboratory activities, receiving more experience and staying competitive in the application of this emerging technology onboard commercial aircrafts is recommended. The activities will include new and even unconventional concepts. This eases the task of selecting the optimum technology in the applications. This applies to all aspects such as harness, topologies, schemes of modulation and protocols. A certain amount of this activities can be performed cost-effectively in the laboratory environment. Continuation is recommended of laboratory activities on the next and more mature components of fibre optic harness, transmitters and receivers; laboratory ageing and testing of components; testing of protocols; and testing of modified protocols.

3.6.2   Subtask 6.2: flight test demonstrations (see Appendix 5.12)

Subtask leader: Fokker Aircraft BV.

The flight test demonstration comprises the installation of a fibre optic network (FON) in accordance with the installation guidelines (Task 3.2) and handling procedures (Task 2.5) into an aircraft, to verify these guidelines and to evaluate installation effects. Besides this, measuring methods as defined in Task 5.4 will be applicated to monitor the performance of the FON. Relevant parameters will be recorded and monitored in flight to show compliance with civil aircraft environmental conditions.

Some of the main items to be decided were the architecture and the components of the FON. This had to be done at a very early stage of the project to be in time for the final flight testing. This meant that because of the time schedule no benefit could be taken from the outcome of other tasks; for example Task 4.2, the selection of first generation of ODT systems. Finally an architecture was chosen with the following main characteristics:

- a STAR network (4 × 4); four transmitter/receivers (Txs/Rxs) each located in a specific aircraft environmental condition representing area (4) which communicate with each other via a transmissive STAR coupler in the cabin of the aircraft

- a long point-to-point fibre link (185 m) running through the four areas (1 Tx/Rx)

- a very short point-to-point link in the cabin of the aircraft (1 Tx/Rx)

- digital data transmission

- NRZ coding

- three types of airborne connectors

- bit error measurement

- optical power level measurement.

The test lay-out has a certain redundancy in terms of utilization, anticipating possible problems with installation, handling, component delivery or others. Relevant parameters (measurands) were measured by the flight test instrumentation system in a Fokker 100 prototype aircraft. That system had to be extended with a data link tester (DLT) generating the digital data for the FON and measuring for each channel the bit errors. Bit errors were defined and counted by the DLT and afterwards implemented in an Arinc 429 data stream which was recorded on a instrumentation recorder. The Txs/Rxs were provided with an analogue peak detector whose outputs were to be measured in-flight, representing the optical power level. Relevant environmental condition parameters were measured as temperatures/pressures in the four areas, temperatures, Txs/Rxs, vibrations bulkhead e.s.o. A goal was to examine the relation between FON performance (as measured by means of bit errors and optical power levels) and aircraft environment.

The intended flight test demonstration was negatively influenced by the bad experience discovered by the needed transceiver/receiver (Txs/Rxs) units. The main problems were as follows.

- too late delivery of the units

- the analogue peak detector output was not as agreed with the manufacturer. The original high frequency pulse of the digital data was not excluded (filtered) in the output signal; signal conditioning of this type of signal was not foreseen; in a last minute change an *ad hoc* electronic circuit was implemented and installed for one fibre optic link

- too low sensitivity of the reciver part of the units; due to this problem the original 6 Tx/Rx configuration FON could not be flight tested; however the structure (fibre harness, Tx/Rx bracketery, flight test instrumentation) was prepared and is ready for the follow up project. FON configuration (No. 1) with the two point-to-point links was tested in ODT flight No. 1 and another

FON configuration (No. 2) with the same links of FON configuration No. 1, a third link with the Tx/Rx in the RH wing and a fourth link with the Tx/Rx in the avionics bay via the STAR coupler in the cabin, were tested in ODT flight No. 2.

The installation of the FON in the aircraft was performed as intended. In total about 400 m FO cable, 12 HA connectors, 2 MIL C38999 connectors (two bulkhead/each four feedthroughs) and eight FSMA connectors were installed.

The ODT flight test demonstration was executed in two flights, each of two hours, as agreed in the Brite/Euram contract. The first flight happened on 6 Dec. 1991; tests in accordance with the testplan were performed. During the whole flight bit error parameters corresponding to the FON configuration No. 1 were monitored and recorded. No bit errors were found. The second flight happened on 18 Dec. 1991, the remaining testplan procedures were finished and some extra take-offs and landings were made. Bit error parameters corresponding to the FON configuration No. 2 were monitored and recorded in flight and the optical power level was measured at the end of the long fibre link. One link whose power budget was marginal showed some bit errors during taxiing of the aircraft. In flight no bit errors were found. The analogue optical power level hardly varied.

Conclusions The installation of a fibre optic network into an aircraft was performed as much as possible in accordance with the guidelines of Subtask 3.2. However it was not fully representative for installation in a new production aircraft, because it had to be added to the existing electrical installation. No serious problems were encountered during the installation, and in spite of concerns for damages to the pre-fabricated harness, they did not occur. The installation test, comprising among others checking the fibre cable harness, showed no faults.

Functional tests on ground and flight tests of the FON, comprising optical data transmission at 500 kbits/s, via different routes through the aircraft running through typical aircraft environmental conditions, showed good operation and no bit errors.

In the first instance hardly any relation could be discovered between aircraft environmental conditions and FON performance; however this point should be investigated more deeply and needs more exposure to flight testing.

No benefit could be taken from outcomes of other ODT tasks in spite of their utility because the flight test demonstration set up had to be frozen while the other tasks were even not started (Subtask 4.2, selection of first generation of ODT systems).

Recommendations Continuation of flight testing with the designed original architecture, getting more exposure to the behaviour of fibre optics in aircraft environment, is recommended. Ageing effects as a concern in this area can be shown and experience can be built up in finding where the problems are and what questions should be asked.

In respect of the above statement, the flight test data processing should search for measured and recorded bit errors. Only in the case of bit errors should other parameters be processed for further analysis.

3.7 Task 7: preparation of draft specifications

The work of standardization carried out by European and other countries constitutes a service in the field of science and technology that is provided for all interested companies in the industrial world, and that confers undoubted benefit on the economy as a whole.

Standardization enables us to create order in our technical world. It is an integral component of the existing economic, social and legal order.

International and regional standards reduce the obstacles to trade and promote world trade. And this makes them particularly important to countries such as shows in the European Union with their heavy reliance on internal and on external trade.

The standards in the highly developed industrialized countries are a source of information on the latest available technology, and are accessible to all. They are invaluable in promoting the worldwide transfer of technology.

Effective standardization is essential to cope with a wide range of technical and economic tasks, it provides all those involved in business and industry with a clearly defined common basis for understanding.

Many protective functions (safety, environmental protection, etc.) would be virtually impossible without standardization.

The use of standardized systems, components, behaviour models and test methods in the development of new airplanes and testing procedures accelerates the introduction of new technical and scientific findings, as well as helping to minimize costs.

Standardized terminology and requirements, quality, safety and test criteria in the European Union and in trade with other countries all help to enhance business on the domestic, European and worldwide level.

This does not imply that the question of most concern to us at the present time can best be solved by as many standards as possible, on the principle that there is safety in numbers. Quite the opposite: it should be endeavoured to restrict new standards to an essential minimum, thus promoting efficiency, continued development of science and technology, as well as international and European harmonization.

Results The main players for standardization in the aerospace technology are the European AECMA (Association Européenne des Costructeurs de Matérial Aérospatial), the US–American AEEC (Airlines Electronic Engineering Committee) with ARINC standards and the US–American SAE (Society of Automotive Engineering). These organizations are in the same way promoters of the standardization activities related to optical data transmission in aircraft.

The AEEC ARINC standards   dominates the commercial market supported by the outstanding US aeronautic industries, the US government and the important US airliners. AEEC is a national US standardization organization; nevertheless ARINC committees and subcommittees are open for international team work. The European aeronautics industry and the European airlines take this opportunity in some committees and subcommittees. For an increase of share in working with ARINC committees, the European parties involved in aeronautics should be stimulated in the future.

The SAE   standardization organization is established in commercial aircraft standardization, as well as in the military and space vehicle side. The SAE international is in charge for working out NATO standards (STANAGs). European NATO partners co-operate as national representatives in ASD committees; however the influence of the European side is comparatively small, e.g. the steering committee of ASD acts without any European member.

Conclusion   The standardization related to fibre optic systems and components in Europe was at first driven by the requirements of military projects. During the running time of the ODT project some European AECMA standards were developed to good draft documents, e.g. the prEN 3910; it was published by AECMA in spring 1992 as a green paper, one of the first fibre optic aircraft standards in Europe.

4   Conclusion

The ODT programme provided a broad field of information about the state of art of optronics. It appears that the selection of an optimum architecture is very difficult for a general hypothetical system. The design guidelines established during the ODT programme could be efficiently applied once a system has been precisely defined. Future studies should include sensor network aspects to have an overall idea of the application of optics on aircraft.

On one hand, this programme has allowed one to identify some open problems which are mainly due to the weak experience of optic in avionics. On the other hand, the results about the feasibility of optical systems allow one to be optimistic about the maturity of the optical technology. An ambitious European demonstration of optical technology performances for aircraft applications is likely to motivate component manufacturers to develop new products well adapted for onboard use.

It may have been too ambitious, considering the timescale and the required interaction between functional work packages (tasks and subtasks) and subconsequently between partners. In fact the lack of sufficient technical discussion between the partners caused a delay in the delivery of documents. Due to this, at the end of the project not enough discussion took place concerning important questions for aircraft manufacturers such as:

• applicability of fibre optics in civil aircraft

• maintainability fibre optics in relation with build in test/monitoring system

• fault tolerance/reliability/certification aspects

• good experience was gained with the practical aspects of fibre optics concerning handling with and installation of it in aircraft

• the flight test demonstration showed good functioning (bit-error-free) of optical data transmission in the aircraft environment

• the tested fibre optic network layout had to be frozen too early in the project to be ready for the final demonstration; in this respect too little feedback existed between the selection of first generation ODT systems and the flight test; on the other hand it is seen that the existing layout is a good start for a following project

• a lot of useful experience was gained during the whole project, working together with the team of European partners; the cooperation grew during the project and the atmosphere was perfect.

The further development of the individual onboard systems in aircraft results in an increase of assignments and functionalities. The number of data required by the onboard systems increases rapidly (doubles every three or four years), the network complexity grows; at the same time stronger real time requirements take place. Therefore high data transmission speed and shorter access time will be inevitable. The answer to this challenge is the introduction of standardized high performance data transmission networks in the next generation of aircraft. Well known problems arising from a copper implementation are solvable in a convincing manner by using fibre optic technology. Wide bandwidth, very high immunity against electromagnetic interference (EMI), perfect disconnection of electrical potential and low weight are outstanding properties of optical fibre. Newer data transmission standards for aircraft use, have consideration for fibre optic implementation, e.g. ARINC 639, ARINC 629 and high speed data bus (HSDB) AS 4074.1 and EN 3910. The next step to advance fibre optics in aircraft should be to gain experiences due to laboratory and flight experiments and to complete standards for systems and components.

5 Appendix

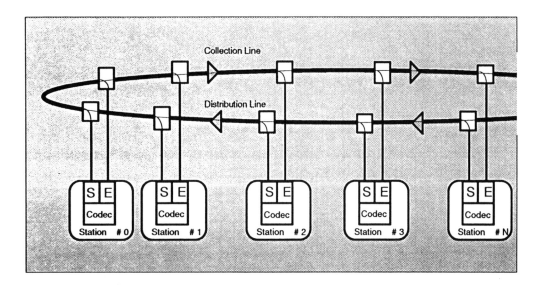

Figure A1   Linear topology

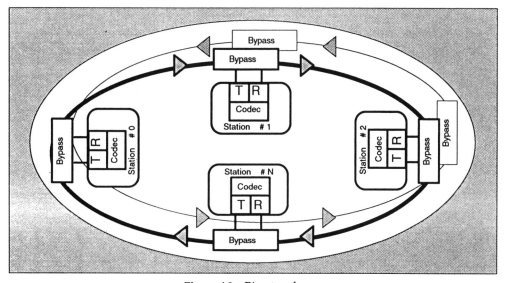

Figure A2   Ring topology

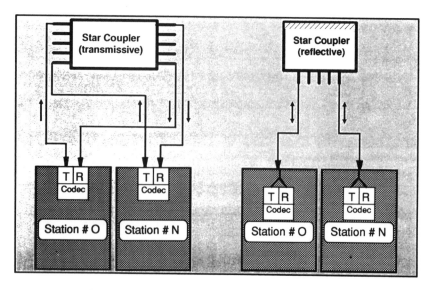

Figure A3 Transmissive and reflective star topology

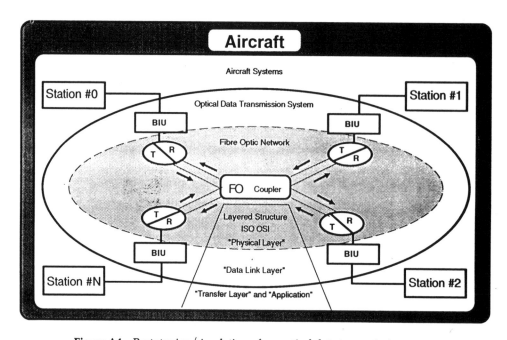

Figure A4 Prototyping/simulation of an optical data transmission system

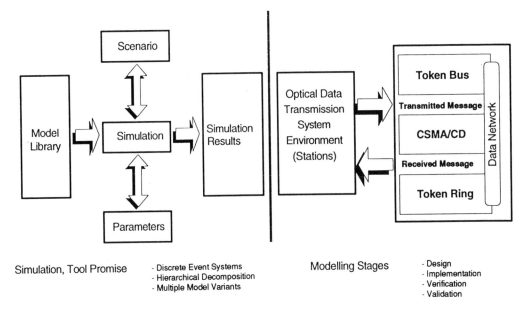

Figure A5 Simulation of an optical data transmission system

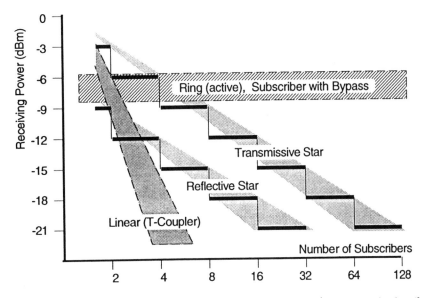

Figure A6 Optical power budget of various network topologies v/s number of subscribers

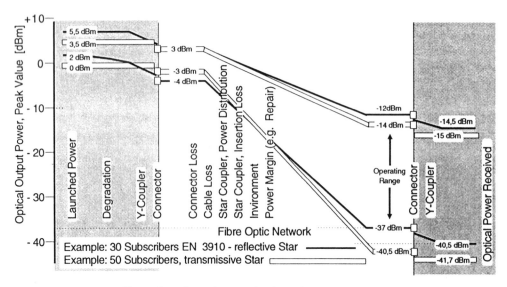

Figure A7 Optical power budget of star topologies

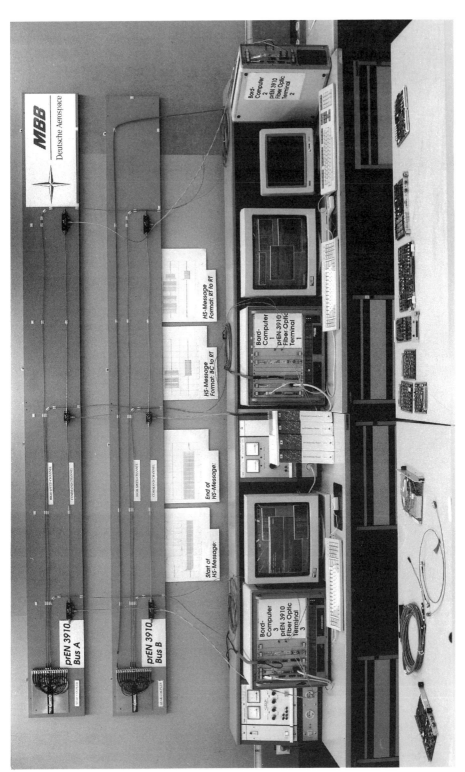

Figure A8 Optical data transmission system, laboratory demonstration

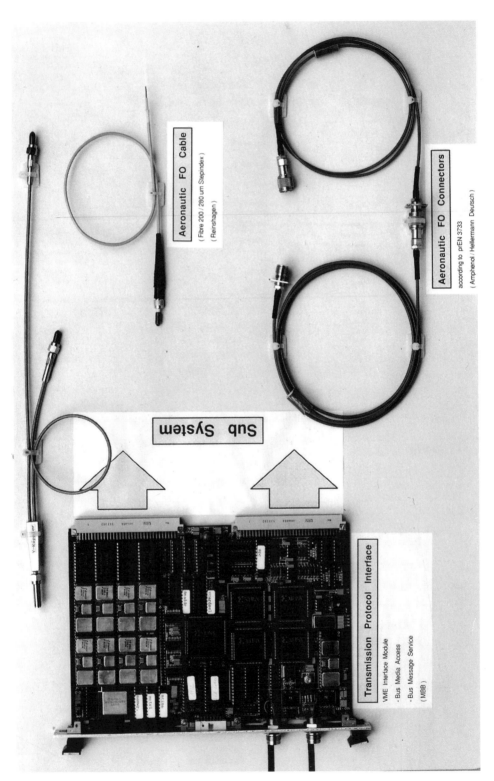

Figure A9 ODT laboratory demonstration, fibre optic components

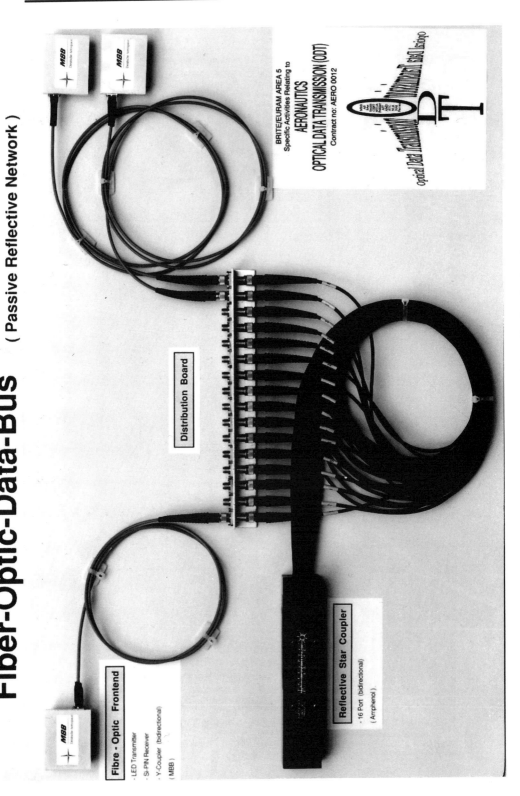

Figure A10 ODT laboratory demonstration, reflective star topology

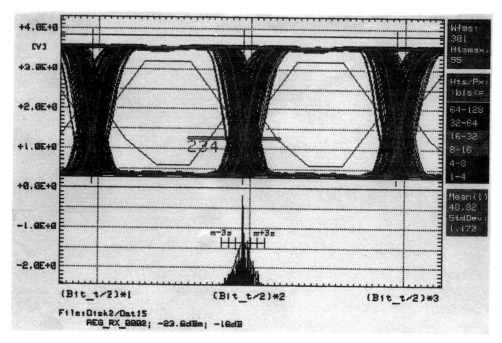

Figure A11   ODT laboratory demonstration, optical signal variation eye diagram

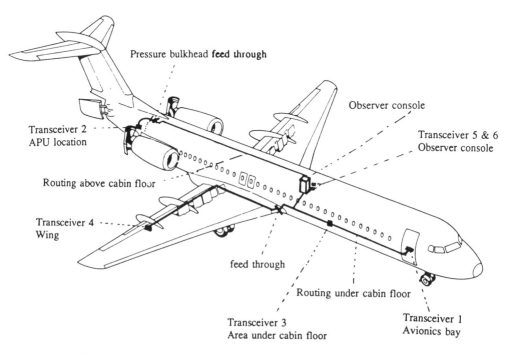

Figure A12   ODT flight test demonstration, installation overview (Fo 100)

2 *New optical sensor concept for aeronautics (NOSCA)*

M. Turpin*

This report, for the period February 1990 to April 1992, covers the activities carried out under the BRITE/EURAM Area 5: 'Aeronautics' Research Contract No. AERO-CT89-0017 (Project: AERO-P1102) between the Commission of the European Communities and the following:

Thomson CSF* (coordinator), France
Aerotecs, France
INESC, Portugal
Sextant Avionique, France
Smiths Industries, United Kingdom
Tu Denmark, Denmark

Contact:
Mr. M. Turpin, Thomson CSF, Domaine de Corbeville, F-91404 Orsay, France
(Tel: +33/1/601970002, Fax: +33/1/60197416)

Advances in Optronics and Avionics Technologies. Edited by M. Garcia
© John Wiley & Sons Ltd.

0   Abstract

This report concerns the NOSCA project and covers the different aspects of the work which has been undertaken during a period of two years. The NOSCA project was split into two stages. The first stage has been devoted to the realization of two separate lines of optical fibre sensors (one for pressure and one for acceleration), using a compatible interfero-polarimetric operating mode. The second stage has been devoted to the evaluation of a complete system including both lines of sensors arranged in the same network and using a common reading and multiplexing system. In addition to the experimental demonstration of such a new optical sensor concept, studies of aircraft require- ments have also been carried out in order to provide consistent guidelines for future developments. Robotic applications of this concept have also been studied. The NOSCA project has been considered as a pilot phase programme, yielding major inputs for a further main phase.

This technical report provides a description of the work undertaken and gives the main results which have been obtained according the workplan defined in the proposal. The results cover the different aspects of the pro- gramme: sensors (pressure and acceleration), networking (light sources, optical links, reading system and processing), modelling, applications (aerospace and industrial (robotics)), and guidelines for future enhanced systems. The report is composed of four scientific and technical sections devoted to the objectives of the project, the research activities, the results and a general conclusion.

1   Introduction

In present avionic systems sensors and sensor data are used in a non- integrated way to serve control functions through dedicated system hardware. New concepts for integrated modular avionics demand a radically new approach for sensors in order to provide a unified system which is compatible with these new architectures. Optical fibre technology provides the basis for a new integrated sensor system concept to meet these requirements.

It is now well known that optical fibre sensors offer many advantages over conventional electrically based sensors in the aerospace environment; these include immunity to EMI, small size, weight, accuracy, dynamic range, and compatibility with optical data buses. Specialized sensors are sometimes required because of the unique requirements and environment encountered in the aerospace industry. It is also to be noted that outside avionics there are parallel technology developments in optical sensing and communication sys- tems to meet the needs of general industrial applications. Another major advantage of fibre optic sensors is the relative ease with which they may be multiplexed into versatile multiparameter networks. Starting from the advan- tages of this promising technology, considering the background of the partners in the field of optics, and being aware of the large international interest in the field of optical fibre sensors and associated techniques for avionics in particular, the NOSCA project was proposed in order to investigate new concepts and provide a broader technology base for future developments.

The NOSCA project, defined as a pilot phase, started in February 1990 for a two year period involving partners from four countries of the EEC: France (Thomson-CSF and Sextant Avionique), United Kingdom (Smiths Industries), Portugal (INESC) and Denmark (DTH).

2   Research objectives

2.1   General

The aim of the project was to demonstrate and provide the basis for an integrated interferometric measurement system with common multiplexing of fibre optic sensors suitable for subsequent development and implementation in avionic systems.

The project has covered the different aspects of sensor networks and used new concepts for intrinsic optical fibre sensors. The sensors have been defined in association with a coherence multiplexing technique and an optical reading system of interferometric type. Two measurands have been selected for the NOSCA demonstrator: pressure and acceleration. For these sensors an innovative transduction arrangement has been proposed and studied. Based on a two stage workplan, the research objectives were mainly the demonstration of the proposed concept of sensors and networking through a series of eight specific tasks:

- study of requirements and definition of specifications

- design of sensor heads, interferometer and processor unit

- development and assembly of sensors

- study of aircraft system criteria for networking

- tests refinement and reports

- assembly evaluation and review of system demonstrator

- guidelines for programme continuation

- studies for robotic applications

The first stage of the project was devoted to the realization of two separate lines of sensors (one for pressure and one for acceleration), assembled in the same serial network arrangement and addressed with the same interfero-polarimetric operating mode. The second stage was devoted to the evaluation of a complete system, including two lines of sensors arranged in the same network and using a common reading system composed of one interferometer and one data processing unit. In addition to the experimental demonstration, studies of aircraft requirements have been carried out in order to provide consistent guidelines for future developments and robotic applications of the sensing concept have been investigated.

2.2   Pressure and acceleration sensors

New concepts of pressure and acceleration measurement based on the intrinsic birefringence properties of specific optical fibres have been proposed. The objective was to demonstrate the capability of such sensors to be compatible with the operational requirements of aeronautics through a demonstrator composed of fully designed and fabricated sensor heads tested in the laboratory environment.

2.3   Optical reading and processing unit

Based on a polarimetric technique, the sensors (pressure and acceleration sensors) produce output optical signals which are coherently encoded. This is mainly due to the use of a broadband light source as a light emitter required for the proposed coherence multiplexing technique. The objective of the work has been to design and fabricate a decoding interferometer in order to provide an exploitable optical signal, and to design and fabricate a processing unit suitable for both pressure and acceleration sensor networks.

2.4   Multiplexing technique

The objective has been to demonstrate that the low coherence multiplexing technique for serial networks composed of several distinct sensors is a powerful mean for sensor addressing and reading, taking into account the parasitic effects induced by the different elements of a network such as optical leads and connectors.

2.5   Modelling

The modelling has been considered as an essential part in the framework of this project in which new concepts for sensors were investigated. It has been performed in terms of detailed study and theoretical characterization of several topics which were relevant for the project implementation. This was the case in the analysis of the accelerometer and pressure sensor heads, as well as in the study of the sensing network: expected performance of the sensors (pressure and acceleration); definition of a useful model of the network; comparison of theoretical models and results.

2.6 Robotic applications

The aim of this section was to study the extension of the sensing system principles in the particular field of robotics. Definitions of sensors requirements, operating ranges, accuracy and reliability have been investigated and the NOSCA sensors performance was evaluated regarding this application field.

3 Research activities

3.1 General

In this section the research activities are described, and they provide the principles of operation of the sensors and the principles of the addressing and multiplexing techniques through five sections: pressure sensor head, acceleration sensor head, measurement techniques, modelling, and robotic applications.

3.2 Pressure sensor head

The specifications for the pressure sensor were reviewed according to the operational requirements. The pressure range of operation, the accuracy and the spurious effect of the temperature were investigated in order to define the specifications of each part of the sensor: optical side-hole fibre for pressure measurement, high birefringent optical fibre for temperature semi-compensation and temperature control, sensor housing including pressure tight inlet/outlet connectors. The pressure sensor has been fully defined and has a serial link of two types of specific optical fibres: a side-hole fibre, devoted to the pressure measurement, and high birefringent 'bow-tie' fibre devoted to the temperature control.

3.2.1 Pressure sensor: side-hole fibre

The pressure sensor is mainly composed of a specific side-hole optical fibre which has a high pressure sensitivity. In this particular type of fibre, two holes are present in the fibre structure, and in this way an external isotropic pressure is transformed into an anisotropic stress in the optical core (Figure 2.1a). The anisotropic modulation of the internal stress allows a variation of the intrinsic

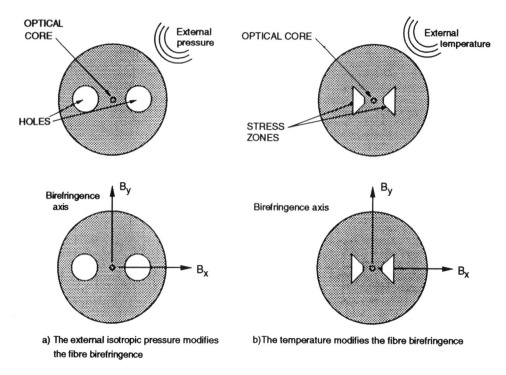

a) The external isotropic pressure modifies the fibre birefringence

b) The temperature modifies the fibre birefringence

Figure 2.1 (a) Side-hole fibre, (b) HiBi fibre.

birefringence of the fibre. The light is launched into the fibre so that both eigen modes are populated and the fibre is equivalent to a polarimeter (or interferometer). The modulation of the internal stress induced by the external pressure changes the propagation constants of each eigen mode and modifies the relative delay (or optical path difference) between the two eigen polarization modes. At the output of a length of fibre, the phase retardation (or pathlength difference) is directly related to the pressure (the spurious effect of the temperature will be discussed after), and can be read by using a coherence decoding system such as a Michelson interferometer.

3.2.2 Temperature compensation: high birefringent fibre

Because the thermal sensitivity of the side-hole fibre is not equal to zero it is necessary to add a specific sensor devoted to the thermal compensation. This sensor is made with an optical fibre which has a low pressure sensitivity compared with the side hole fibre and a high temperature sensitivity. This optical fibre has an intrinsic thermal induced stress birefringence which is directly related to the temperature (Figure 2.1b).

3.2.3   Pressure sensor head arrangement and design

The pressure sensor head is defined by a combination of two lengths of optical fibres, spliced together in order to give a single serial local network arrangement (Figure 2.2). The first fibre is a side-hole fibre (pressure detection with a low dependance versus temperature) and the second fibre is a stress induced birefringence fibre: a bow-tie fibre. By this way the sensor provides both pressure and temperature information. The design of the pressure sensor housing has been made to satisfy the main requirements of environmental conditions (temperature, aeronautic fluids . . .) and also in terms of size compared with standard sensors. A typical sensor is composed of the following elements: 10 m of side hole fibre, 2 m of high birefringence fibre, and a specific housing with a standard pressure inlet and pressure tight inlet/outlet for the optical fibre up and down leads.

3.3   Acceleration sensor

3.3.1   Acceleration sensor arrangement and design

The acceleration sensor is based upon the inertial mass principle. The sensor consists of a fibre wound on a compliant cylinder which is loaded by the inertial mass (Figure 2.3). A positive acceleration acts to axially compress the cylinder resulting in a radial expansion, the fibre thus experiences an effective axial strain. A double headed configuration is used to allow temperature compensation schemes to be implemented. The design of the acceleration sensor head has encompassed both a theoretical and an experimental analysis. The theoretical model developed predicts that the major contributions to the stiffness, which determines the sensor sensitivity, of the sensor coils is due to the restraining effect of the optical fibre and that the cylinder material has only

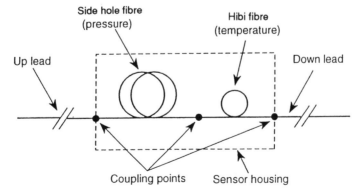

Figure 2.2   Pressure sensor, schematic arrangement.

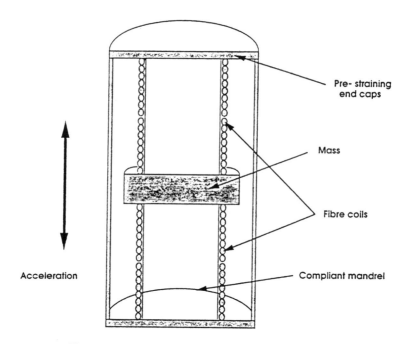

Figure 2.3   Acceleration sensor, schematic arrangement.

second order effect. A detailed investigated into the optical and mechanical component properties was performed in order to optimize the sensor components and configuration.

3.4   Measurement techniques

This section covers three major aspects of the NOSCA concept:

- the optical reading system: interferometer unit
- the data capture and treatment: processing unit
- the multiplexing technique: interfero-polarimetry

3.4.1   Interferometer unit

Because the information provided by the sensors is optically encoded a specific optical receiver is required for the decoding operation. After decoding, the information can be exploited by the processing unit. A Michelson interferometer has been chosen as the optical decoding receiver. The Michelson interferometer is a two-beam interferometer which has two separate arms. One of the arms has a fixed length providing a reference optical path. The second arm is arranged in such a way that its length can be modified. The input light is separated into the two arms, reflected at mirrors and combined into a common output path before reaching the detector. In such a device, the optical path difference is given by $\delta = 2nL$, where L is the geometrical path difference and n is the refractive index of the medium (air in this case).

The interference visibility function reaches a maximum value when both arms of the interferometer are perfectly balanced and when both waves are perfectly coherent. For a given interferometer and a given light source there is in all cases a zero optical path difference which provides the maximum output signal. This particular point corresponds to a position of the mobile mirror which defines the 'zero' of the device and acts as the reference point when an absolute phase delay is required. Starting from this zero-point, interference fringes are obtained when the mobile mirror is displaced. The visibility of the interference fringes is directly related to the light source coherence. The interferometric measurements, the value of the unknown variable is obtained from the phase delay which is deduced from the number of interference fringes. Figure 2.4 shows an example of theoretical interferogram. The Michelson interferometer has been defined in order to be compatible with both pressure and acceleration sensors. Specific requirements such as optical path range, interferometric displacement control of the mobile mirror and optical stability have been taken into account. The output signal transmission has been studied in order to optimize the interfacing with the processor.

3.4.2 Processor unit

A detailed definition and functional block diagram of the processor unit was developed. The processor unit's input and output requirements were specified, and the process unit/interferometer interface was defined to ensure compatibility with both accelerometer and pressure sensor signals. A modular approach to the design was adopted to allow for ease of implementation and to permit design modifications to be made easily. A schematic block diagram shows the electronic functional arrangement of the processor in Figure 2.5. A standard rack sensor was used to house the modular electronic boards. The rank system is self contained with its own mains driven power supply and backplane. Interconnections to the interferometer (optical signals and electronic control signals) SLD drive and PC are provided by connectors on the front panels. The processor unit consisted of the following circuit boards: two SLD drives, detector/demodulator, HeNe detector/decision logic and memory, interferometer controller, PC interface.

3.4.3 Multiplexing technique

The aim of NOSCA was to demonstrate the potential of optical fibre sensors (pressure and acceleration) and the capability of such sensors to be optically

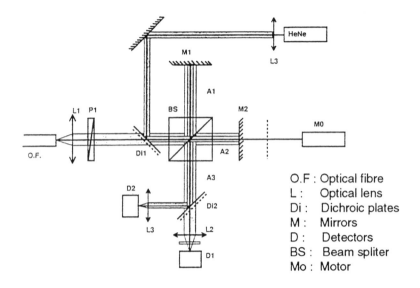

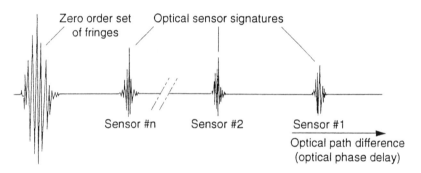

Figure 2.4   Michelson interferometer and interferogram definition.

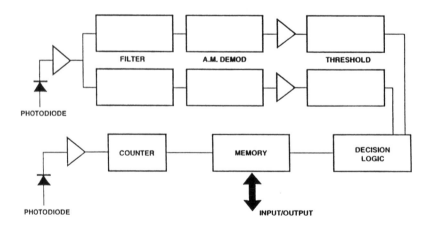

Figure 2.5   Processing unit, functional diagram.

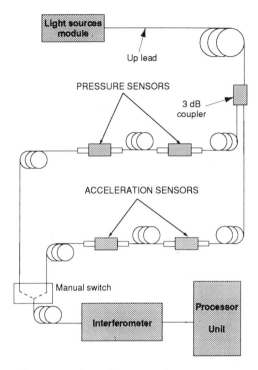

Figure 2.6 Complete network arrangement.

multiplexed. The principle of the NOSCA multiplexing technique is based on the coherence properties of light in polarimetric and interferometric systems.

Polarimetric scheme The polarimetric concept is based on the birefringence properties of specific optical fibres. High birefringence can be used in two different ways which are implemented in the polarimetric sensor networks. Starting from a broadband light source, insensitive optical fibre leads are obtained by launching light into one of the eigen modes of the fibre. The high birefringence allows one to conserve an input linear state of polarization over a long length of fibre. The polarimeter, which is the sensing part of the network is obtained when both eigen modes are excited. As a consequence of the birefringence, a differential propagation of both eigen modes is seen, yielding a phase delay which can be measured. The transfer of light from one axis to the other is obtained by a controlled local rotation of the eigen axes of the fibre (or a controlled splice if two different fibres are used), and the amount of coupled light is related to the relative axes rotation. The end of the sensor network is defined by a 45° coupling point, which allows one to use the output fibre as an insensitive lead. This arrangement provides a well defined remote sensing element localized on an optical fibre serial network and can be extended to several sensing parts, providing a multi-sensor serial network.

Broadband light source   The low temporal coherence properties of a broadband light source maximize the advantage of the coherence multiplexing technique. Considering a primary wavetrain emitted by the light source, each sensor generates two uncorrelated secondary wavetrains which cannot overlap (because the time delay introduced by the polarimeter is greater than the coherence time of the light). In this way a plurality of sensors can be arranged so that a plurality of non-overlapping wavetrains is generated. The phase delays of the sensors are contained in a channelled spectrum which is the physical result of the wavetrains' superposition.

Decoding and processing   The output channelled spectrum contains all the phase information but is not directly in the right form for processing. In order to obtain an exploitable signal the decoding interferometer is used. As the mobile mirror of the interferometer is moved, the optical path difference is scanned from zero to a value which corresponds to the maximum optical path delay provided by the sensor network. The fundamental laws of the broadband interferometer produce an interferogram which corresponds to the phase mapping of the network. A typical interferogram is composed of a plurality of fringe patterns, each corresponding to a sensor signature (Figure 2.4). The optical fringe patterns are in a suitable form for processing. The role of the processor is to identify the fringe patterns, to determine the corresponding phase delay and to recover the pressure and/or acceleration values for each sensor in the network. The related research activity has concerned the assembly of the pressure and acceleration network and the integration of the measurement system composed of the interferometer and the processor.

3.5 Modelling

The modelling research activity has been conducted in two different correlated fundamental aspects: modelling of sensors (pressure and acceleration), and modelling of the multiplexing scheme.

3.5.1 Study and model sensor head

Models have been developed, based on fundamental principles, to evaluate the basic sensitivity of the pressure and acceleration sensors. In terms of pressure sensitivity, the side-hole fibre structure has been investigated in order to evaluate the stress distribution obtained in the structure for different size of circular holes and to optimize the structure. The temperature sensitivity of the side-hole fibre has also been modelled. In terms of acceleration sensitivity, the proposed sensor arrangement (fibre wrapped with tension around a compliant cylinder) has been studied, in order to optimize the transduction coefficient.

For both applications, models have been developed from the Theory of Elasticity and classical mathematical tools.

3.5.2 Study of the system performance

The interfero-polarimetric principle based on the low coherence multiplexing has been investigated from a theoretical point of view. A model of the proposed serial sensor network has been developed in order to evaluate the sensor performances. A model of the effect of the coupling points in terms of signal level, associated noise, second and higher order crosstalk effects and power budget has been carried out. The aim of this systematic theoretical approach was to determine the different sources of noise created by the system, to evaluate each of them and to provide the basic rules for the implementation of optimized networks: optimized coupling ratio for each sensor, maximum number of sensor possible and signal-to-noise optimization.

3.6 Robotic applications

A complete overview of the potential applications of fibre optics sensors into robotics areas has been carried out. Measurands of interest have been identified, pressures such as accuracy, bandwidth, drift, hysteresis, linearity, etc . . ., have been investigated in order to provide the specific requirements for sensors into this area. Other important parameters such as size, weight and environmental conditions have also been taken into account.

4 Research results

4.1 General

This section presents the main results which have been obtained during the project through the theoretical and experimental investigation and implementation of the NOSCA demonstrator. It concerns the following topics: pressure sensors, acceleration sensors, optical decoder, data handling and processing, networking and multiplexing, theoretical research and modelling, and robotics applications.

4.2 Pressure sensors

The design of the pressure sensor housing has been made in order to satisfy the main requirements of environmental conditions (temperature, aeronautic fluids . . .) and size compared to standard sensors. The sensing element of the sensor is composed of two lengths of optical fibre: typically 10 m of side hole fibre, and typically 2 m of high birefringence fibre.

The two fibres are spliced together with a small local rotation of their neutral axes in order to get a coupling point of polarization used to separate the pressure and temperature information. Starting from a computer generated design and drawings (Figure 2.7), the pressure sensor heads have been fabricated and assembled (Figure 2.8). The two fibres (including the splice) are wound together in the sensor housing and specific stuffing boxes have been designed to ensure tight optical fibre input and output. Several pressure sensor heads have been assembled, and basic characterization in terms of pressure and temperature has been carried out. As an example, one intrinsic configuration of the pressure sensors is detailed here, including length and type of fibres used, and the comparison of thermal and pressure sensitivities before and after sensor assembly.

up lead:	2.4 m of HiBi fibre
pressure sensor:	10 m of side-hole fibre
temperature sensor:	2.5 m of HiBi fibre
down lead:	5 m of HiBi fibre

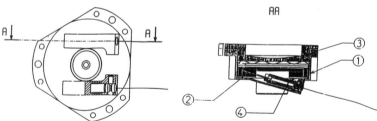

Figure 2.7   Pressure sensor design.

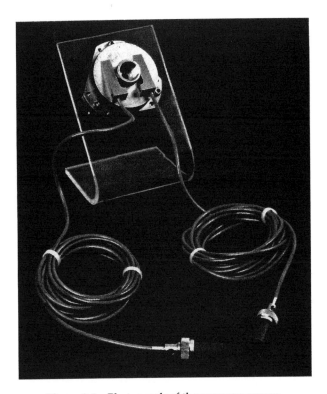

Figure 2.8   Photograph of the pressure sensor.

The different fibres of the sensor head have been spliced in order to obtain coupling points of about 20% in intensity, except the last coupling point (output) which induces an intensity coupling of 40%. This sensor has been tested in terms of temperature and pressure sensitivities, as described below.

Over the range $-10\,°C$ to $40\,°C$ the measured phase shift induced by 10 m of the side hole fibre is -5.1 rad/°C, which corresponds to a temperature sensitivity of

$$St = -0.51 \text{ rad/°Cm}$$

The measured phase shift for the temperature sensor in 12.25 rad/°C, obtained with 2.5 m of HiBi fibre, which corresponds to a temperature sensitivity of

$$St = -4.9 \text{ rad/°Cm}$$

The pressure sensitivity has been measured in the range of 1–10 bars with a specific test bench which is mainly composed of an hydraulic sensor (pump, tank, . . .). The measured pressure sensitivity of the side hole fibre is deduced from the phase shift of -34 rad/bar obtained with 10 m of fibre:

$$Sp = -3.4 \text{ rad/bar m}$$

In the same range of pressure (1–10 bars) the pressure sensitivity with 2.4 m of HiBi fibre is about 2.64 rad/bar which corresponds to

$$Sp = 1.1 \text{ rad/bar m}$$

The characteristic response curves of the sensor are shown in Figure 2.9–2.11 which respectively give the pressure sensitivity of the side-hole (FASE) fibre for three different temperatures: 0 °C, 21 °C and 40 °C, the pressure sensitivity of the HiBi (bow tie) fibre at the same temperatures, and the sensor head output

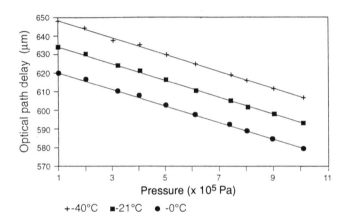

Figure 2.9   Pressure sensitivity of the side-hole fibre.

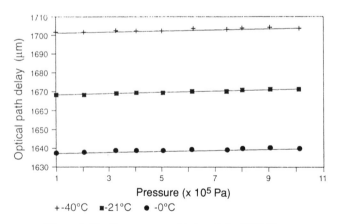

Figure 2.10   Pressure sensitivity of the HiBi fibre.

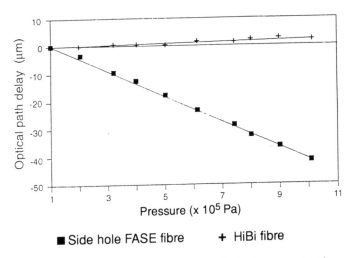

Figure 2.11 Compared pressure sensitivity (sensor output).

comparison of responses of the side-hole and HiBi fibres for a pressure range of 1–10 bar at room temperature.

Additional investigations have been carried out in terms of pressure and temperature with different pressure sensor heads. As an example, a sensor head composed of 10 m of another side-hole fibre has been tested including environmental tests. The main results are give here.

Configuration of the sensor

10 m of side-hole fibre
pressure sensitivity: 1.4 rad/bar m
temperature sensitivity: -0.43 rad/°Cm

Calibration tests

Optical path difference (at 20 °C/1 bar): 800 μm
interferogram: correct (no spurious coupling points)
pressure calibration:

range   1–11 bars
temperature   20 °C
sensitivity   14 rad/bar
linearity error: $+/-$ 2%

Temperature calibration:

range	−20 °C, 40 °C
pressure	6 bars
sensitivity	−4,3 rad/°C
linearity error:	+/− 2%

Environmental tests results

Over pressure (20 bars)	no damage
vibrations (standard aeronautics tests)	no damage
shock (12 g)	no damage
humidity (10 days, RH 95%, 65 °C)	no damage

4.3 Acceleration sensors

A schematic of the accelerometer design is shown in Figure 2.12. Two fibre-wound mandrels are placed on axes either side of an inertial mass; this arrangement is held securely within a skeleton frame. A number of parameters may be varied to optimize the design; these include: mass, fibre length, and mandrel nature and dimensions. The main design goal was that an acceleration

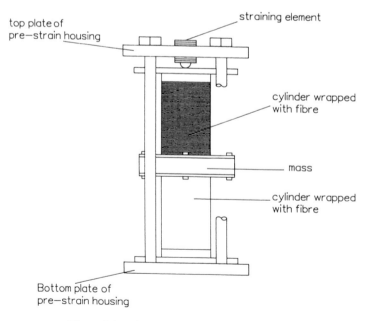

Figure 2.12   Schematic of the accelerometer.

range of $+/-$ 10 g induces a 100 fringes output range previously specified. To aid with the design progress, a theoretical model was developed. Input data concerning fibre sensitivity to an axial strain, for the model was obtained experimentally and is shown below for a number of different high birefringent commercially available fibres.

Fibre types and manufacturer	Beat length (mm)	Extension required for one fringe (μm)
York Bow tie 125 μm	1	50
York Bow tie 80 μm	1	50
3M EOTEC	2	120
Andrews 'E' series	1	900

The York bow tie 80 μm clad fibre was chosen because of its high axial strain sensitivity. In the wound configuration the 80 μm fibre shows increased sensitivity over the more usual 125 μm fibre. The fibre is also specifically designed for sensor applications in which the fibre is tightly coiled. A photograph of the accelerometer is shown in Figure 2.13.

The results of the modelling showed that the mandrel material played only a secondary role, and so a highly stable aerospace rubber compound was chosen. Care was taken in winding the fibre coils to ensure that an even coverage was obtained and therefore maximum stress was transferred to the fibre and that there were no twist which may introduce spurious coupling points. The sensor coils were also pre-stressed to allow both positive and negative accelerations to be measured. The model predicted a sensitivity of about 5 fringes per g, which agrees well with the experimental values ranging from 4.2 to 5.5 fringes per g. Figure 2.14 shows a typical calibration curve; the acceleration output is linear within experimental error.

The sensor head is configured with two sensor coils to enable temperature compensation schemes to be implemented. The two coils may be detected separately and the compensation performed electronically. This has the added advantage of reducing the sensor path length difference and therefore the travel required in the decoding interferometer.

4.4 Measurement techniques

4.4.1 Interferometer unit

The function of the interferometer is to deliver an interferometric signal exploitable by the processor by decoding the encoded information of optical path differences (i.e. phase delays) generated at the output of the sensors or at the output of the network. A Michelson-type interferometer has been built

Figure 2.13   Photograph of the accelerometer.

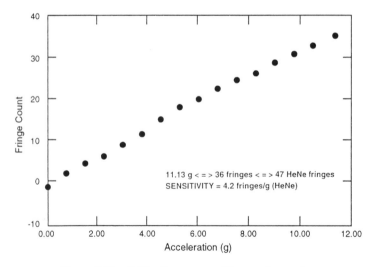

Figure 2.14   Calibration curve of the accelerometer.

with bulk optics components. A specificity of the NOSCA interferometer is the use of a highly coherent laser (HeNe) emitting at 633 nm providing interference fringes of high visibility for all the positions of the mobile mirror. In this way, the absolute value of the displacement (optical path difference) can be determined within an accuracy with plus or minus one half-fringe. In parallel of this interferometric calibration system, the light coming from the sensors is launched into the interferometer. After passing through a collimating lens and a polarizer, the light beam is split in to two equal secondary beams. The first beam propagates following a fixed optical path (reference arm of the interferometer) and the second beam propagates in the adjustable arm, obtained by the controlled displacement of a mirror attached on a translation stage driven by a DC motor. At the output, the two beams are recombined and they overlap with a phase difference which is a function of the optical path difference introduced in the interferometer. Two output optical fibre links have also been added in order to feed the detectors (one for the signal and the other for the HeNe reference) attached to the processor. A photograph of the interferometer is shown in Figure 2.15.

4.4.2 Processor unit

A photography of the processor is shown in Figure 2.16. The function of the processor unit is to detect and demodulate the sensor and reference signals

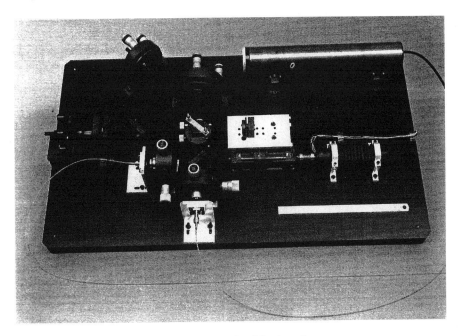

Figure 2.15 Photograph of the interferometer.

Figure 2.16 Photograph of the processor unit.

generated in the interferometer. The sensor signal is a sinusoidally varying interference pattern within a gaussian envelope. The pattern is generated in the interferometer by linearly displacing the mirror. The position of the mirror is referenced to the balance point of the interferometer and is measured using a coherent source and fringe counting techniques. The requirement is to determine the exact position of the central fringe of the pattern with respect to the mirror displacement. Simple peak detection of the envelope is not sufficiently accurate because of the shape of the gaussian envelope and the level of noise present in the small signals. A number of alternative signal processing schemes were investigated and a trade-off study performed. The dual wavelength fringe identification scheme chosen works on the principle that when the signal and decoding interferometers are exactly matched there is always constructive interference (i.e. the two patterns are in phase) irrespective at the wavelength of the illuminating source. However, if there is a slight imbalance the two fringe patterns appear out of phase. By using two SLDs of slightly different wavelengths, the phase matching of the two patterns may be detected and hence the central fringe determined. Sub-carrier techniques are used to separate the two SLD signals; one SLD is amplitude modulated at 100 kHz and the second is modulated at 450 kHz. Detected signal are shown in Figure 2.17. The two fringe patterns are recovered by filtering the respective signals. The signals are then thresholded, squared up and fed to a logic circuit which determines the phase match to within one fringe. This is used to trigger a memory which records the relative fringe count of the HeNe fringes.

4.4.3   Multiplexing technique and network assembly

The pressure sensor network has been successfully assembled and demonstrated through a number of modified arrangements. One complete network composed of two non-packaged sensors has been assembled with up, intermediate and down optical leads of 10 m length but without connectors. For this first approach, splices (fused and/or glued) have been used. A second network has been assembled with the complete packaged sensor head, optical leads and connectors. Specialized Radiall connectors have been used in order to control the orientation of the polarization axes and consequently the energy coupling ratio. For both networks, the configuration optimization has been performed by adjusting the main relevant parameters such as weak coupling points and power budget.

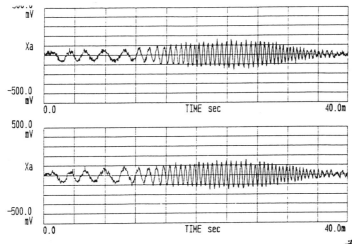

Upper trace: signal from SLD 1
Lower trace: signal from SLD 2

Figure 2.17   Detected output signals (SLD 1 and 2).

The acceleration sensor network has been also successfully assembled and demonstrated. A number of network configurations have been investigated and the optimum configuration has been implemented for the demonstration.

4.5 Complete network results

A complete network has been assembled and tested in terms of compatibility by using an arrangement (see Figure 2.16) composed of the followings sub-systems:

- light sources emitter subsystem

- pressure sensor network

- acceleration sensor network

- interferometer subsystem

- processing unit subsystem

4.5.1   Light sources emitter subsystem

Two superluminescent diodes with polarization-maintaining pigtails provided by Thomson-CSF have been selected and linked together by using a 3 dB polarization maintaining coupler. By this way the two slightly different mean wavelength of the SLDs (about 5 nm offset) have been combined and were available at each output port of the coupler with a mean optical power available of about 150 μW each.

4.5.2   Pressure sensor network

Fully assembled with optical connectors, one representative interferogram of the network is shown in Figure 2.18. In order to compare the effect of networking on the sensor signature visibility, this figure shows together sensor signatures and sensor network signatures for pressure sensors. It can be seen that the network assembly does not strongly affect the sensor signature visibility.

4.5.3   Acceleration sensor network

The final system was fully connected using the special Radiall HiBi fibre connectors. The connectors were found to be non-ideal, however they did make the assembly of the sensor easier and more representative of a practical

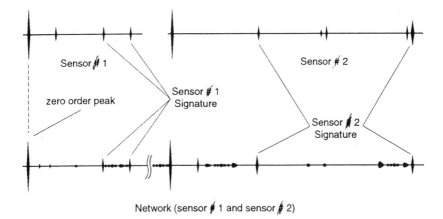

Network (sensor # 1 and sensor # 2)

Figure 2.18   Interferogram of the pressure sensor network.

implementation. A typical interferogram for this network is shown in Figure 2.19; the largest signal is the interferometer balance point, the next signal is due to the second sensor head and occurs at the path length difference associated solely with the second sensor. The next signal occurs at the combined path of the first and second sensor. Each sensor signal represents the sensor difference terms because of the strong coupling.

4.5.4   Interferometer and processor

The interferometer and the processor have been integrated in order to provide a complete system for the network data capture. A personal computer is used

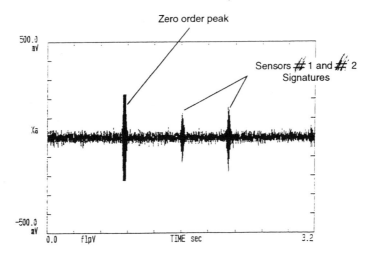

Figure 2.19   Interferogram of the acceleration sensor network.

for calculations and data edition. As an example a complete scan of the acceleration sensor network is shown in Figure 2.20 in association with the corresponding digitalized signal provided by the processor.

4.6 Comparison between modelling and results

Theoretical models have been developed, based on basic principles, to evaluate the basic sensitivities of pressure sensor and accelerometers. A network analysis has also been performed in order to evaluate and quantify the spurious cross-interference effects and to determine the best arrangement for the proposed serial network suitable for low crosstalk effects and enhanced signal-to-noise ratio.

4.6.1 Sensor models

The method used for calculation of the pressure sensitivity of the side-hole fibre is based on complex potential theory applied in two-dimensional elastic problems. The temperature sensitivity of this type of fibre has also been calculated using the Theory of Elasticity and the thermoelastic displacement potential functions. Different configurations of the side-hole fibre have been investigated, taking into account parameters such as: size of holes, distance between holes and diameter of the fibre. It should be noted that for all cases, circular holes have been considered.

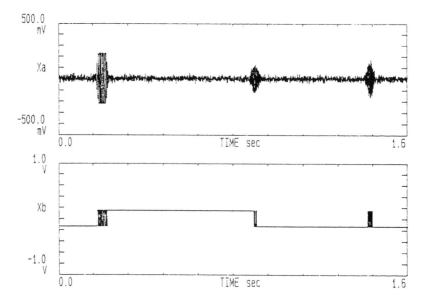

Figure 2.20   Network interferogram and digitalized fringe systems.

Similar methods have been used for calculation of the stress induced bire-fringence in HiBi fibres such as Panda and bow tie fibre used for the measurement of acceleration. For the temperature sensitivity, parameters such as thermal expansion coefficients have been used for the calculation. These parameters depend strongly on dopant materials and the contents that are employed for the silica glass. Generally these parameters are not well known.

In terms of acceleration sensitivity, highly birefringent fibres have been considered. In this case, because the fibre is wrapped around a compliant cylinder, a phenomenological model has been developed to estimate the sensor sensitivity. Comparisons between theoretical model and experimental results are summarized below.

Pressure sensor

Side-hole fibre	Theoretical	Experimental
Pressure sensitivity	−3.9 rad/bar m	−3.4 rad/bar m
Temperature sensitivity	−0.7 rad/°Cm	−0.5 rad/°Cm
Highly birefringent fibre		
Temperature sensitivity	−2.7 rad/°Cm	−4.9 rad/°Cm

Acceleration sensor

Highly birefringent fibre	Theoretical	Experimental
Acceleration sensitivity	5 fringes/g	4 or 5 fringes/g

It can be seen that the results are in good agreement. Nevertheless the slight differences which can be observed are well explained: in the case of the side-hole fibre the pressure sensitivity can vary from one fibre to another. This is mainly due to the size of the holes and to the distance which separates the holes from the core of the fibre. Figure 2.21 shows the pressure sensitivity and the temperature sensitivity of the side-hole fibre as a function of the structure dimensions. In the case of the HiBi fibre the pressure sensitivity is strongly related to the nature of the stress zones. The rheological properties of the material which constitute these zones are not provided by the supplier. Nevertheless the order of magnitude given by the model is correct, compared to the experimental results. Figure 2.22 shows the temperature sensitivity of the HiBi bow tie and panda fibres.

For acceleration sensitivity, the calculated value as the experimental value is strongly dependent on the nature of the mandrel. It can be noted that the

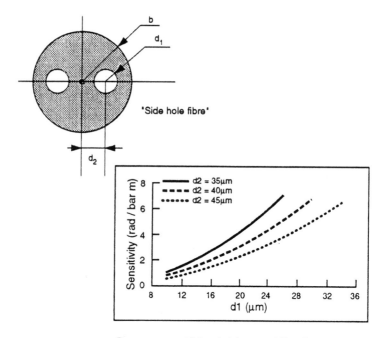

Pressure sensitivity of side-tunnel fibre for b = 95μm

Figure 2.21   Theoretical side-hole fibre pressure sensitivity.

model provides a numerical value in near-perfect agreement with the experimental result.

4.6.2   Network analysis

Starting from the proposed concept of coherence multiplexing, the serial network arrangement has been studied in terms of performance, taking into account the major parameters such as coupling point level (weak and strong ratio of coupled energy), path imbalances of the interferometers (polarimeter) and minimum detectable phase signal (signal-to-noise ratio). Considering the analysis relative to crosstalk between sensors for the situation corresponding to the NOSCA arrangement:

UP LEAD – SENSOR 1 – INTER. LEAD – SENSOR 2 – DOWN LEAD

it has been found that when the sensors have exactly the same path imbalance, the crosstalk (expressed as the ratio between the spurious higher order signal level and the sensor signal) is equal to a quantity directly proportional to the coupling ratio of energy k, and equal to 10 k. In practice this value is lower and gives a signal-over-noise ratio of 5. In fact the level of crosstalk is smaller than

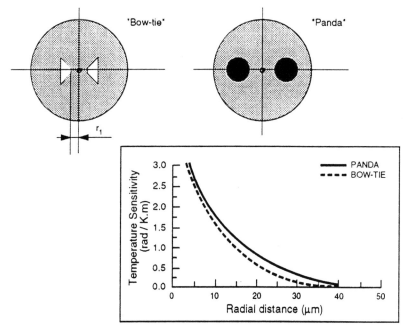

Temperature sensitivity of PANDA and BOW-TIE fibres as a function of r_1

Figure 2.22 Theoretical HiBi fibre temperature sensitivity.

predicted because the theory is obtained with the assumption that path imbalances of the sensors are equal, which is not the case in practice. Taking into account this fact, it can be noted that the model developed for the analysis of the network crosstalk is consistent with the experimental results obtained. For both pressure sensor and acceleration sensor networks, relatively weak coupling points have been made for the sensors and leads discrimination, according to the rules provided by the theoretical model analysis. In practice a coupling point is realized by a local rotation of the eigen modes of a fibre by a rotated splice between two fibres. The angles of rotation realized in practice are typically about 15°, which corresponds to a coupling ratio of energy of about 0.07. This small value corresponds to the particular case of weak coupling points. As predicted by the theory, the crosstalk level in this case is lower than the noise floor of the sensor (source noise, photon noise . . .).

4.7 Robotic applications

Sensors used in robots are specified with the same performance parameters as those used for sensors in other applications. An evaluation of the performance parameters for the NOSCA sensors as internal sensors for robots is given in the next paragraphs.

4.7.1 Absolute versus relative measurement

Absolute sensors are generally preferable to relative ones for robotic applications. If relative sensors are used in an industrial environment, the problem of 'missed counts' can be different to solve. Both of the NOSCA sensors are absolute sensors.

4.7.2 Accuracy

This is the degree to which the actual location corresponds to a desired or commanded location. Resolution and hysteresis (see below) are often much more important than accuracy. The accuracy claims of robot suppliers are often in fact the repeatability. The repeatability is often better than the actual accuracy. The NOSCA sensors, in the present version, have an accuracy of +/− 2% F.S. This is four times too much for robotics applications since the demands for the robot sensors are +/− 0.5% F.S.

4.7.3 Bandwidth

The bandwidth of the sensors should be at least as great as that of the joint servos. In fact, sensor bandwidths greater than 20 times the servo bandwidth are often required. The robot servo bandwidths are normally up to 100 Hz. The demand for the bandwidth is more than 200 Hz for the accelerometer, and more than 2 kHz for the pressure sensor.

4.7.4 Drift

Drift can be a special problem in force and torque sensors. These values must often be measured accurately in the region near zero. In this case, a small drift in the sensor output can result in unacceptable errors.

4.7.5 Hysteresis

Excessive hysteresis in robot sensors can be drastic. Hysteresis is often larger than the sensor resolution, in which case the useful resolution is essentially equal to the hysteresis. Hysteresis in a servo system can make an otherwise stable system become unstable. Often the only acceptable solution to these instabilities lies in eliminating the hysteresis. The hysteresis is typically 0.1% F.S. or better for robot pressure sensors, the same as for the present NOSCA pressure sensor.

4.7.6 Linearity

This is the degree to which the slope of the input/output curve is constant, i.e. the curve is linear. A sensor should introduce a minimum of unlinearity to the servo system. The linearity of the present NOSCA sensors is approximately 2% F.S., and the demand for robot sensors is 0.1%, or less, which is more than a factor ten too much for the NOSCA sensors studied in the project.

4.7.7 Monotonicity

Sensors with digital outputs (and occasionally analogue sensors) can have non-monotonic input/output relationship (i.e. the slope of the input/output curve changes sign). Such sensors cannot be recommended for use in robots because they almost invariably cause servo instability. The NOSCA sensors are strictly monotone.

4.7.8 Noise

Noise can be an unexpected problem in robot sensors, especially analogue ones. Like hysteresis, excessive noise can decrease a sensor's useful resolution.

4.7.9 Range

This is the extent to which or the limits between which variation is possible to measure by the sensor. The range of a sensor must, of course, include the working range of the robot for a given physical size to be measured. The requirements for robot pressure sensors are in the range 0–200 bar, the range of the NOSCA pressure sensor is 0–10 bar; this is more than a factor ten too small. The range for robot accelerometers is 0–2 g, and 0–10 g for the NOSCA accelerometer; this is five times too much.

4.7.10 Repeatability

This is the closeness of agreement of repeated movements in size (position, pressure, force, etc.), under the same conditions, to the same reference value (position, pressure, force, etc.). A repeatability of less than +/− 0.1% F.S. should be reasonable for most sensors.

4.7.11 Resolution

A measure of the smallest possible increment of change in the variable output of a device. This is often the most important factor and it is five times too large.

4.7.12 Sampling rate

Sampling rate for a sensor is the frequency at which the sensor value (or values) is read. This means that the sensor value is read at discrete time intervals. Based on the value read, an actuator command is computed and sent to the actuator. The sampling rate for the NOSCA sensors will not exceed 10 Hz; this is at least five times too slow for robotics applications, since the absolute minimum sampling rate for robot sensors is 50 Hz (200 Hz may be required for the pressure sensor if possible).

As seen from this evaluation, the NOSCA sensors (as studied in the project) do not completely meet the requirements for performance of robot sensors, at least with respect to: accuracy, linearity, range, resolution, response time (bandwidth), and sampling rate. Furthermore, the weight and probably the size of the present NOSCA sensors are larger than the currently available sensors of the same type for robotics applications; these mechanical properties of the sensors are critical, especially for smaller types of robot.

4.8 Discussion of results

4.8.1 Pressure sensor

As defined in the project proposal, the pressure sensor heads have been assembled and tested. The results obtained are in good agreement with the expected performances in terms of sensitivity and accuracy. Tested in the range 1–10 bars, the sensors present a response which has good linearity. Tests have been performed for different temperatures in the range −20 °C to 40 °C and no degradations of the sensor performance have been observed. In order to obtain a tight sensor housing, specific tight inlets have been implemented, and good operation of the sensor head has been successfully verified for pressure of up to 20 bars. Additional tests such as moisture tests, vibrations and shocks have also been performed without any observed drastic effects on the sensors.

For the interconnection, commercially available polarization maintaining connectors have been used. These connectors are obviously not optimized for this particular application and consequently the problems due to interconnection have not been optimized. This implies some extra losses which reduce the power budget of the system. Nevertheless, in order to demonstrate the potential performances achievable with this type of connector, one specific sensor head has been assembled with connectors directly attached to the sensor housing. The pressure sensor network concept, based on the assembly of polarimetric sensors, has been implemented and, according to the theoretical models, the different optical parts of the network have been assembled by splices and connectors with the creation of weak coupling points of energy. Such an assembly has provided results in good agreement with models: the signatures (fringe patterns) can be easily identified on the resulting interferogram and can be detected without ambiguity by the processor unit after decoding by the interferometer. No degradation of the signal has been detected after networking, even if spurious coupling points appear on the interferogram, due to interconnet effects.

4.8.2 Acceleration sensor

Acceleration sensors have been assembled and tested according to the scheme defined in the proposal. A double headed design has been adopted based on a push-pull arrangement, avoiding the effect of temperature which acts on the two coils of the sensor in a similar fashion. The performances of the acceleration sensors have been compared with models and are in good agreement. As for the pressure sensors, weak coupling points have been made between the sensitive fibres and connectors to ensure low crosstalk. The sensor housing has been designed in order to provide a sensor suitable for laboratory tests, and they must not be considered as an optimize device.

In terms of network, the acceleration sensors have been assembled with the same serial arrangement that the pressure sensors. Up, intermediate and down leads have been interconnected with polarization-maintaining connectors. By using this method, the two sensor networks (pressure and acceleration) have been made fully compatible. The observation of the acceleration sensor signatures shows that the fringe patterns have a good visibility, can be easily detected and, as with the pressure sensor network, no degradation of the signal has been observed after networking of the accelerometers.

4.8.3 General remark

The performances obtained within the NOSCA project must be considered as preliminary results obtained under laboratory conditions.

The side-hole fibre used in the frame of the project is not fully optimized in terms of high pressure sensitivity and low temperature sensitivity. Nevertheless side-hole fibres with high pressure sensitivity up to 12 rad/bar m have been obtained and could be advantageously used for enhanced sensors. Considering the temperature range, the limitation is mainly due to the classical epoxy-acrylate coating of the fibre. It is well known that high temperature coatings (up to 350 °C) are available and could be used with the side-hole fibre and the HiBi bow tie fibre for both pressure and acceleration sensors.

The performances obtained in terms of network are in perfect agreement with the theoretical models. However it is to be noted that in the case of large networks, with more than five sensors, the serial architecture does not appear to be the optimum choice. In this case different basic architectures could be investigated, such as, for example, ladder, tree . . ., or hybrid topologies such as ladder and serial, tree and serial . . .

4.8.4 Interferometer and processor unit

The interferometer and the processor unit have been built in order to provide laboratory means for optical decoding, data handling and processing. Concerning the interferometer, the goal was not to fabricate an integrated device and bulk optical components have been used. The performances of the interferometer met the requirements and were compatible with the processing unit. It is to be noted that multimode optical fibres have been used to transmit the optical signal to the processor and this original arrangement has been successfully validated.

The processor unit has been built according to the general processing scheme for interference fringes detection. In addition to the classical electronic cards (modulation, demodulation, triggering, . . .) an original method of fringe detection has been proposed and implemented. Based on the use of two light sources modulated at different frequencies and having a slight wavelength difference, this new method has been validated successfully. As for sensors, these subsystems have not been optimized regarding the requirements for real avionics environment.

4.9 Feasibility

The feasibility of an innovative optical fibre sensor network has been demonstrated through theoretical and experimental results in terms of sensor principles and multiplexing.

4.9.1 Basic points

The following principles, based on the use of optical fibre as the sensing element, have been demonstrated.

- Intrinsic optical fibre sensors (pressure sensors and acceleration sensors) have been built and successfully tested in laboratory environment.

- Compatibility of both types of sensors has been demonstrated by using the same addressing and reading technique.

- Multiplexing, based on the broadband interferometry principles in association with the polarimetric properties of the sensors, has been investigated and demonstrated.

4.9.2 System

Associated to the sensors, a complete system has been assembled and tested with success.

- An optical decoding interferometer has been fabricated, providing exploitable signals compatible with the data handling.

- A processing unit has been developed, fabricated and used for the data handling and processing.

4.9.3 Networking

For networking, the key components have been installed and tested.

- Insensitive optical fibre links have been installed and the remote operation of the sensors has been demonstrated.

- Connectors have been used for the assembly of the network.

Starting from the basic demonstration carried out by the end of the NOSCA project, the feasibility of the proposed concept has been made at the laboratory level and further developments are required for the implementation of such sensor networks in the aeronautics environment.

4.9.4 Areas requiring further investigation and development

The area of single mode fibre components is only now being addressed and, as with multimode fibre, it is the telecommunication market which is the driving force.

- Specific HiBi fibres are required for sensing: these fibres are in their experimental stage; however, different types of these fibres are now commercially available, and more specific fibres such as pressure sensitive side-hole fibres can be easily fabricated with enhanced sensitivity for sensor applications. Specific coatings for the high temperature operating range are also now available, and by this way the use of optical fibres in operational systems can be envisaged with a minimum of risks. It is not only the fibre components which require development but also the opto-electronic components. A wide range of devices is available, but they are rarely optimized for sensor applications. Nevertheless, optical light sources and detectors are available and can be used, and the field of integrated optics offers considerable potential for components. Integrated optics offers the opportunity to integrate a large number of functions on a single chip. A number of specific components which require development have been identified.

- Connectors: low loss single mode and low loss single mode polarization-maintaining connectors are required to successfully implement the proposed networks in a practical form.

- Modulators: phase modulators are required and can be obtained in a specific form starting from the existing devices.

- Polarizors: fibre polarizing components, couplers, depolarizors and polarization controllers are required to enable all-fibre systems to be implemented in a practical form. Such components are now available and can be adapted.

- Sources: low coherent light sources are required with low Fabry–Perot modulation. Such components are now available but higher power is required.

- Cabling: fibre protection schemes are required to enable the sensors developed to be successfully transferred from the laboratory.

After having demonstrated the NOSCA concept in the laboratory, the design for a compact interferometer needs to be investigated and experimental models produced. Practical implementations may impose further design criteria on both the sensors and networks and, as such, this is seen as a key area of work for the future. It is also an essential prerequisite to progress of the work into the aircraft environment.

Signal processing techniques have been demonstrated. It is anticipated, however, that in future implementations a more sophisticated adaptable scheme might be required. The new or adapted design will have to be compatible with the function of the compact interferometer, and should also take into account the aircraft sensor system requirements, both functional and environmental.

To progress out of the laboratory and on the aircraft, fibre systems will be required to meet the full environmental specifications. To meet this, not only ruggedized fibre components are needed but also robust systems. The development of fibre optic components is very much driven by the communication market; not only are specialized components required to perform functions in aerospace applications but they are also required to operate in harsh environments. Testing procedures need to be developed for qualification testing. Installation issues need to be addressed preferably via mounting the sensor interconnect, etc ... in an aircraft mock-up. This will not allow the impact of the installation procedures, installation space envelope, network routing, etc ... on the sensor performance to be addressed, but would also provide valuable experience for installation engineers.

Networking encompasses not only the addressing, architecture and interrogation scheme, but must also take into acount such factors as routing and interfacing to avionic systems. Many of these issues are application-specific, however the basic ground work may be performed for generalized situations. The choice of network architecture in this pilot phase was made on the basis of achievability with the available components. The network has a number of drawbacks and it is recommended that alternative networks are investigated, which are functionally robust and more compatible with aerospace requirements.

Standardization activities should form a vital part of any future programme. It is recommended that in collaboration with the aircraft manufacturers, an early start is made to form the framework for optical sensor systems standard in aircraft.

5 Conclusion

The NOSCA project started in February 1990 as a pilot phase programme for a period of two years.

The aim of the project was the investigation of intrinsic fibre optic sensors. A new concept for fibre optic sensor was proposed and measurands (pressure and acceleration) were chosen for the demonstration. In addition an innovative addressing scheme based on the fundamental laws of broadband interferometry was proposed for the multiplexing and networking demonstration.

The different sub-systems of the demonstrator have been studied, designed, assembled and tested in order to provide a complete set of experimental results suitable for an exhaustive and comprehensive analysis of the capabilities of such a sensor concept. Pressure and acceleration sensor heads have been built and tested in a laboratory environment. Based on the same principle of operation (the polarimetric principle) both types of sensor have been characterized in terms of sensitivity and accuracy but also in terms of practical arrangement (housing, size, weight . . .).

According to the technical objectives of the project, the pressure sensors have been tested in the range 1–11 bar and the expected resolution of 0.1 bar has been achieved. In terms of temperature the pressure sensors have been tested in the limited range $-20\,°C$ to $40\,°C$. In this range, the good operation of the pressure sensor has been demonstrated. Similarly the acceleration sensors have been tested in the range $-10\,g$ to $10\,g$ with the expected resolution of $0.2\,g$ within the same temperature range. This limited temperature range is mainly due to the coating of the optical fibres and is not to be considered as an intrinsic limitation of the sensors. The use of specific coatings such as poly-imide can easily solve this problem.

In terms of networking, two separate networks have been assembled in a first step. One network with two pressure sensors and one network with two accelerometers. For each of these, a complete interconnection has been made by using insensitive highly birefringent fibre optic leads and specific polarization-maintaining connectors. The implementation of such networks with commercially available components has demonstrated the network feasibility. In a second step, both networks have been linked together and addressed with common multiplexing elements. In order to demonstrate the multiplexing sub-systems such as light source modules, interferometer and processor have been assembled and tested. Because the aim of the project was mainly a laboratory demonstration of the concept, the interferometer has been built in a laboratory form (not suitable for avionic environment). Nevertheless, solutions based on opto-electronic devices such as integrated optics can be developed, starting from existing components in order to obtain a compact and robust solid state system. In the same way, the processor unit has also been built for a laboratory demonstration and could be easily refined to match the aeronautic requirements.

In terms of industrial applications of the NOSCA concept, a review of specifications for robotics area has been carried out and guidelines for such applications have been provided.

In performing the work for NOSCA, observations concerning the current state of fibre optic technology, components, addressing and handling techniques have been made, and areas which require further basic research or development have been identified. In addition to further development to undertake in the fields of fibres, passive components such as connectors, and active components such as integrated optics, specific work has to be performed in terms of network topologies, ruggedization of sub-systems and standardization.

Based on the main results obtained in the frame of the project, the NOSCA project has demonstrated the feasibility of constructing compatible new types of optical fibre sensors for different measurands based on a single technique, and also that these sensors may be integrated onto a single network using common multiplexing elements.

In this pilot phase project NOSCA, it has been demonstrated that pressure and acceleration sensors may be implemented on parallel networks with a single decoding interferometer and processor unit. It is important to note that similar techniques may be applied to a large number of measurands: linear position, proximity, temperature, speed, flow, fuel quantity, . . .

6   Acknowledgements

The author would like to acknowledge the partners for their contribution to the work carried out during the NOSCA project: M. Brevignon, J.P. Le Pesant

(Thomson-CSF), R.M. Taylor, M.J. Ranshaw (Smiths Ind.), O. Gaouditz, M. Breda, H. Leblond, J.P. Batisse (Sextant Av.), J.L. Santos, A.P. Leite (INESC), F. Conrad, E. Trostman (DTH), and R. Leclercq (Aerotec).

7 List of publications related to the NOSCA project

M. TURPIN, New Optical Sensor Concept for Aeronautics NOSCA, Aeronautics days proceedings, April 16 and 17 1991, Brussels.

R.M. TAYLOR and M.J. RANSHAW, Coherence multiplexed polarimetric fibre sensor arrays for aerospace application, *Optics and Lasers in Engineering*, **16** (1992), 223–236.

M. TURPIN, M. BREVIGNON, J.P. LE PESANT and O. GAOUDITZ, Interfero-polarimetric fiber optic sensor for both pressure and temperature measurement, *OFS'8 conference proceedings*, January 29–31 1992, Monterey.

J.L. SANTOS, A.P. LEITE, Multiplexing of polarimetric sensors addressed in coherence, *OFS'9 conference proceedings*, May 4–6 1993, Firenze.

3 Integrated modular avionics general executive software (IMAGES)

D. Graves, R. Meunier and P. Aldegheri*

This report, for the period February 1990 to January 1992, covers the activities carried out under the BRITE/EURAM Area 5: 'Aeronautics' Research Contract No. AERO-CT89-0015 (Project: AERO-P1033) between the Commission of the European Communities and the following:

Aerospatiale* (coordinator), France
Alenia, Italy
Alsys, France
British Aerospace Airbus, United Kingdom
Captec, Ireland
CRI, Denmark
Deutsche Aerospace, Germany
Deutsche Aerospace Airbus, Germany
Fokker, Netherlands
INESC, Portugal
INSAT, France
NLR, Netherlands
NTU Athens, Greece
SAS, Belgium
SCYT, Spain
Sextant Avionique, France

Contact:
Mr. D. Graves, Aérospatiale, Dept. A/BTE/SYI, Route de Bayonne 316, F-31060 Toulouse, France
(Tel: +33/61/936461, Fax: +33/61/938090)

0 Abstract

The aim of this project was to perform studies on the Integrated Modular Avionics (IMA) concept relative to the definition of the functional specifications of the Executive Environment Software.

This project was focussed on basic software for processing modules, namely core modules which are able to support several application software units developed with the Ada programming language.

The main objectives of this project were to define firstly the Executive functional specification and secondly the Executive/Applications interface specification in order to influence the different ARINC committees in the standardization activities of the IMA interfaces.

For this purpose, special attention was paid to:

- the requirements of the Executive Software

- the use of Ada programming language

- the certification and integration aspects

- the communication issues

- the envisageable distributed system architectures.

The presentation of the results of the study of these topics constitutes the main part of this document, of which the conclusion lists the major points which need further study in order to complete the approach to the IMA concept.

1 Introduction

The major goal of cost reduction for the next generation of aircraft induces us to design Integrated Modular Avionics (IMA), intended to reduce the amount of avionics equipment by a factor of ten.

This is intended to reduce the weight and volume of embedded equipment, by sharing the system resources, and to obtain a greater flexibility for the inevitable evolution in this area.

In order to reach this goal, it is proposed to provide complete independence between embedded applications and non-specific powerful hardware, by using common standardized Executive Software.

This study is intended to provide the European aircraft industry with a thorough knowledge of the resident Executive Software functions. As the studies are covered by a large part of the European aircraft industries, this leads to common understanding of the requirements.

Several European countries are represented by one or more companies in performing the IMAGES project, and the main aircraft manufacturers have a big involvement in the project.

This document describes: what has been performed in the IMAGES project, and what kind of further work is required to obtain a detailed specification of the Executive Software.

This document is divided into three parts: the first part presents the research objectives of the IMAGES project; the second part presents all the research results according to main topics; and as a conclusion, the third part deals with the various main points which it is necessary to study in order to have a thorough knowledge of the IMA concept implications.

2 Research objectives

The aim of the IMAGES project was to provide functional specifications of the Executive Software, which has to:

- give all the systems services for multiple-application and module management purposes
- ensure robust insulation between the various applications which run on the same module
- provide a standardized interface with the application software (APEX)

So, such a project is required for European companies to participate efficiently in the definition of an ARINC standard which would define such Executive Software.

Moreover, the increase of competitiveness in the development of avionics can only be achieved by a greater independence between hardware and software and this is a great challenge for European aeronautical industries.

This study will provide the aeronautical industries with the opportunity to increase their knowledge of new software technologies, through the collaboration with industries and universities that are expert in Ada technology.

3 Research results

3.1 General

The results of this study are split into five topics:

- use of the Ada programming language
- development and certification process
- communication between applications
- distributed system architecture

3.2 Executive Software definition and APEX interface

3.2.1 General description

The IMA concept consists in having a set of shared resources available for supporting the various embedded avionics applications. These standardized resources are based on multiple-user buses (ARINC 629) and cabinets (described through ARINC 651 paper). Each cabinet includes line replaceable modules in order to provide resources such as power supply, processing, I/O and means for internal communication (ARINC 659 backplane bus).

The aim of the IMAGES project was to define standardized Executive Environment Software, allowing the integration of several avionic applications software units inside the same processing module. In order to ensure complete independent between hardware resources and software, it is necessary to define the functions of this Executive Software.

3.2.2 Function description

The Executive Environment Software is the basic software of the processing (or core) module. Its role is both to support the processing of multiple applications and to manage the basic functionalities of the module (e.g. downloading and module initialization). The piece of software which deals with the multiple applications aspects is named Executive Software.

In order to make it possible to achieve interchangeability of Ada application

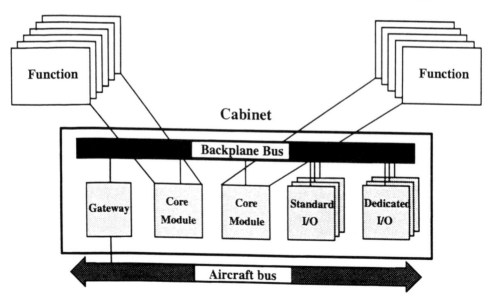

Figure 3.1

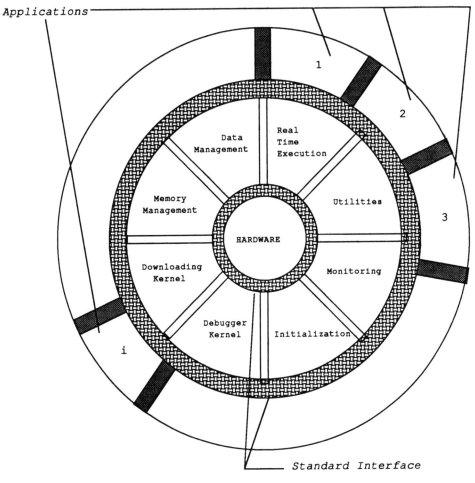

Figure 3.2

programs, a standardized interface (APEX) is defined. Another interface (COEX) exists between the Executive Environment Software and the hardware. This latter interface is outside the scope of the IMAGES project.

Before describing the various functionalities of the Executive Environment Software, it is worthwhile to give the general principles followed to meet segregation objectives.

- Partitioning: there are no direct links between applications. Any interaction between applications must use services provided by the Executive Software through the APEX interface. This rule is applicable to any relation between applications and Executive Software.

- Module behaviour: The various functionalities of the Executive Environment Software are strictly scheduled which makes it possible to define processing modes. For instance, module initialization must be over before activation of the multiple application execution (i.e. Executive Software) mode. Moreover, this latter mode and the downloading mode are exclusive and their activation depends on environmental conditions. In the same way, all the debugging kernel means cannot be used at any time.

Real-time execution   The real-time function is the part of the Executive Software which manages the real-time execution of the applications. So the problem raised consists in defining the link between the application scheduling and the application process scheduling.

For segregation and simplication reasons, it has been decided to define two levels.

- One execution time slice is defined for each application. The scheduling of this time slice is controlled by the Application Dispatcher.

- Within each time slice, the processes of one application are scheduled.

As the Ada programming language is used to implement the application software; basically, the task scheduling is under the control of the Ada RTS. The assumption supposed in IMAGES is that the Ada RTS is part of the Executive Software and is shared between the applications of the module. In order to ensure transportability of the application software, standardization of the interface between Ada RTS and different Ada compilers used for application coding is necessary.

Communication   In the IMA context, it is interesting to suppose, as application programs are independent of the hardware implementation, that they do not know their actual location.

Thus, the Executive Software is in charge of all the physical transmission issues (message building, routing, redundancy management). In particular, it is required that the same request be used by the application to communicate within the core module, with another module of the cabinet and with the outside of the cabinet.

Segregation constraints impose that all the application requests for communication only use the services of the APEX interface. Both directed (point-to-point) and broadcast transmission must be envisaged.

Furthermore, hard real-time requirements demand simplication of the seven layer OSI model (especially periodic transmissions) in order to become efficient.

The IMAGES project proposes to define a distributed data base which covers the whole cabinet, and both ensures the consistency of data exchanged between applications and minimizes the transmissions latency. This data base will have to manage correctly the physical transmission paths for all the communications of the cabinet.

Complementary studies have shown that the communication philosophy has a big impact on the Executive Software implementation and that the nature of the backplane bus (and thus its standardization through the ARINC 659 group) is an essential parameter.

Health monitoring An implication of the integrated module avionics (IMA) concept is the use of a distributed system architecture. Distributed computing on aircraft requires reliable, high-availability and high-performance Executive Environment Software that is capable of coping with the real-time constraints on aircraft, while supporting the execution of applications.

The maintenance philosophy for IMA systems is to have a deferred maintenance interval. For this purpose, the system shall be fault-tolerant.

Fault tolerance is the ability of equipment to provide its function and to continue its operation in a defined manner after one or more faults. This implies the following system design requirements:

- redundancy management

- fault detection

- fault handling

In order to support system fault tolerance, the software shall be capable of reconfiguration.

Reconfiguration in a distributed environment is a powerful mechanism that will enhance flexibility, increase useful lifetime and facilitate maintenance.

In the IMA concept Software, reconfiguration consists in the redistribution of redundant application software modules among the different remaining non-faulty resources.

For dynamic reconfiguration, which implies activation of new applications in an operational system, special attention must be given to the following constraints:

- memory

- run time

- real-time

To control optimal reconfiguration, detailed monitoring of the state of the system needs to be provided.

In a distributed environment, system monitoring is performed on different hierarchical levels.

Local monitoring is performed by the local Executive Software, resident on each core module via the health monitoring function. This health monitoring function must manage three kinds of failure:

- the ones resulting from the hardware resources which are detected by built-in test (BIT) means

- the ones linked to the monitoring of the Executive processing itself (access violation, communication failures)

- the ones detected by the applications and which may be provoked by functional causes independently of any local resource failures; this means that the APEX must also provide services for communication between applications and the health monitoring function.

Global monitoring is performed by a special executive function, called the supervisor. This special executive function consists of redundant supervisor processes, which run parallel and are monitored.

To meet reliability requirements a minimum level of redundancy of active supervisor processes must be maintained. The supervisor process provides the following functions:

- global state monitoring by communication with local health monitors

- system fault log in system database

- global reconfiguration control by dictating rescheduling of applications

Redundant system state data can be organized in a system state database. By using a distributed database it is possible to set up a control structure, based on the same hierarchical structure as for monitoring.

Such a database should be high reliable and available and offer transparency to the application programmer.

It is obvious that such a supervisor function has a privileged place to communicate with the onboard maintenance system and that it is worthwhile to ask for a new global approach to maintenance purposes.

Debugging kernel description The high throughput of current processors makes the use of external tools for software debugging and testing more difficult every day. It is a good reason for the Executive Environment Software to have a resident debugging kernel. So this debugging kernel should be well suited to Ada programs.

The purpose of the study was to describe an architecture model supporting

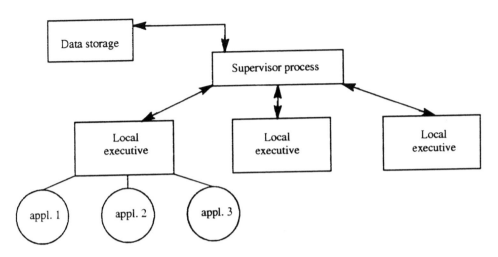

Figure 3.3

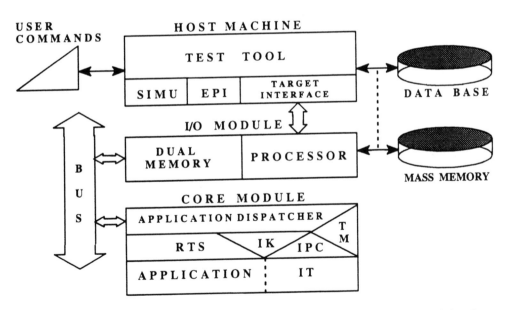

RTS   : Run Time System   IK   : Instrumentation Kernel   EPI   : Emulator Programmatic Interface

IPC   : Inter Process Channel IT   : Instrumentation Task   SIMU : Simulator

TM   : Target Monitor

Figure 3.4

the different tools used during debugging or testing phases. A description follows.

The set of tools, namely simulator, emulator and debugging kernel, are merged within a unique tool, in order to have a common user interface and an Ada oriented tool, knowing Ada Run Time characteristics and implementation.

This unique test tool may work under different modes.

- Simulated mode: this mode is activated when using the simulator. The simulator can be used very early in the development process, either with a real target or an emulator. Its usage is more valuable in testing and evaluating limited parts of the software than when debugging it as a whole.

- Emulated mode: this mode is activated when using an emulator. The great advantage of using an emulator is that it is not intrusive. The program will behave exactly like reality.

- Cross mode: this mode is activated when using the target monitor for debugging. A target monitor is used when the microprocessor is based on RISC architecture or integrating multiple caches and pipelines. In this case it is not possible to use an emulator. The target monitor communicates with the test tool via the target interface. So a set of commands is necessary to implement the target monitor.

- Embedded mode: this mode is activated when validation and flight tests are performed.

In order to get information from the run-time system, a new component called the instrumentation kernel (IK) has been defined. The access to the services provided by the IK introduces an overhead which must be time-bounded. A set of services is provided by the IK.

In order to get information from the application, a new component has been defined, namely the instrumentation task (IT). The IT is a part of the application and is in charge of collecting requests from the user or the application itself, performing the request either directly or by calling the IK and returning the result of the request to the application or the user.

In order to implement all the services provided by the IK, it is necessary to have an appropriate run-time system. The task control block must contain specific information, the different queues used by the run-time system must be designed with special instrumentation, a scheduling mechanism must be also instrumented in order to record scheduling events such as task switches or to provide a background process required for CPU load measurement.

Downloading kernel description   The downloading function has to integrate application updates and configure the IMA cabinet accordingly.

Typically, using a host machine, an operator will be able to modify the application codes and configuration tables in the implied modules in order to implement a new avionics state safely and without any removal of modules.

It is a very critical function of the Executive Environment Software because it does not respect the basic segregation rules formerly described since it can access and modify non-volatile memory areas. For this reason, when downloading takes place, the real time context of the Executive Environment Software is disturbed, and then it is necessary to activate the downloading function only when very particular environmental conditions are present. It seems normal to link this activation to time when the plane is on park station and special downloading requests have been made.

The basic role of the downloading function consists in:

- managing the communication with the host machine

- checking the suitability of the version of loaded applications between themselves and with the Executive Environment Software

- copying the application codes and configuration tables in a non-volatile memory and checking that programming is correct.

I/O interrupt handling In order to design a core module with one or more processors that is able to run multiple applications, the processor on the module has to meet several requirements. These requirements are caused by the demands made upon the IMA system; e.g. fault tolerance, multi-applications and multi-tasking. For this purpose the processor needs to be equipped with exception processing capabilities. In modern processors a wide variety of exceptions is provided. For the IMA system these are evaluated and divided into two groups: necessary and desirable exceptions. Necessary exceptions are regarded as being vital to the operation of the IMA system, while desirable exceptions are considered to be useful.

The study covers both the hardware- and software-generated exceptions and their demands imposed on the backplane bus. The exceptions that are generated by hardware are called interrupts and are regarded as a subset of all processor exceptions. Necessary interrupt signals are the bus error and power-up reset.

A bus error is generated when an application tries to access a location in a memory which it does not have the privilege to, or when data transfer was unsuccessful. The power-up reset signal is generated by power modules when a cabinet is powered up. After a reset, a built-in test will be performed and all applications in the cabinet will be started. Both interrupt signals need to be transferred via the backplane bus to the core module(s).

Desirable interrupt signals are AC failure and system failure. The AC failure signal is generated by power modules as soon as they detect an interrupt in the AC supply, to enable storage of critical data in non-volatile memory before the DC power drops. The system failure interrupt can be used in conjunction with both intelligent and non-intelligent modules. After a reset, intelligent modules enable the system failure signal and perform a self-test. After a successful self-test they disable the system failure signal. In the case of non-intelligent modules, the core module runs the test. Also, intelligent modules can enable the system failure signal when they detect an internal error during operation. Both AC and system failure signals need to be transferred via the backplane bus to the core module(s).

Other exceptions do not make demands upon the backplane bus; they are either generated on the core module or inside the processor. Because core modules communicate with each other over the backplane bus, the bus should provide certain facilities to communicate both interrupt signals and periodic or aperiodic data.

Tasks can be interrupted by events that force the processor to execute an exception or interrupt service routine. After this exception or interrupt service routine has been processed, another task may have a higher priority than the one that was interrupted. Therefore, the IMA system has to be provided with a mechanism to enable task re-scheduling. Scheduling and exception processing having to be interfaced by software.

Because typical IMA applications should have a deterministic behaviour, periodic I/O is likely to predominate. Periodic device handlers can be executed after each application or after a complete cycle of all applications. Both schemes imply poor bus utilization. SAFEbus is a design concept for the ARINC 659 backplane bus with better bus utilization. Bus time is divided into fixed windows and the cycle of operations is table-driven.

There is also a possibility for aperiodic I/O. The Ada *rendez-vous* mechanism supports aperiodic I/O by linking low-level hardware interrupts to high-level Ada code.

Initialization   The initialization phase is usually used to perform special treatments and tests (BIT) required before passing into the real-time mode. Initialization can occur on the ground or in flight.

It is necessary to separate the initializations into two kinds.

- Module initializations are split into: power loss (they can be short, medium and long according to well defined durations) and Executive Reset due to external or internal resets.

- Application Resets are split into: external events (such as action on push button), internal events forced through the Executive Software (health monitoring function), and internal events forced through the application itself.

3.2.3   Operating system survey

Seven existing real-time kernels are reviewed according the functions of the Executive Software. The following topics are addressed: task, memory, interrupt and time management, as well as communication and several miscellaneous features, such as development tools and performance. The kernels reviewed for this study are: ARTX/VRTX32, pSOS⁺, VxWorks, OS-9, the Spring kernel, CHORUS and Alpha.

VRTX, pSOS⁺, VxWorks and OS-9, are commercially available and functionally alike. For instance, the scheduling algorithm employed by all is pre-emptive priority-based.

Spring, CHORUS and Alpha are of a distributed architecture. As opposed to the commercially available kernels, they are able to support multiple applications. They are equipped with sophisticated scheduling algorithms, such as deadlines or so called time-value functions.

Neither of the systems studied and discussed for this sub-task is directly suitable to be used for IMA. Although they offer the possibility to design deterministic application software, the commercially available kernels and their add-ons are not able to support multiple applications. The support of multiple applications is an important requirement of the executive.

Spring, CHORUS and Alpha are able to support multiple applications, but they are not deterministic and thus not easily certifiable.

3.2.4 Executive/application interface (APEX)

Introduction   The aim of this study was to present and specify the definition of the Executive/application interface (APEX).

The following description considers the four fields studied in the IMAGES project:

- definition of requirements of such an interface

- consideration of Ada extensions required for avionics applications

- Ada specification of the APEX services

- use of these APEX services for an application example

Executive/applications interface requirements   The study is based on the assessment of current representative equipments performed by four airframe manufacturers. It has surveyed the main characteristics of current embedded applications and defines the frontier between applications features and Executive ones.

Four steps can be identified.

Application function identification: using the studies performed by the involved partners, the most important features provided by embedded applications are listed and described. It appears that three modes of functioning must be noted: initialization, operational (or basic) and failure modes. Regarding functionalities, it is necessary to split them between real-time execution functions, communications functions, avionic computations, health monitoring, maintenance facilities and flight tests means. The need for equipment redundancy also imposes some changeover logic.

Survey of the functions provided by the Executive Software: using concurrent results concerning the Executive Software functional requirements, it can be concluded that to have a standardized environment for the applications is similar to having all the functions as real-time execution, communication and module health monitoring provided by the Executive.

Definition of the frontier between application and Executive Software: a concrete distribution of the different features between Executive and applications is proposed. The identification of needs for an interface deals with the fact that some items of the provided functions must be shared by the applications and the Executive.

Description of the service requirements: the IMAGES project assumes that primitives of Ada RTS and extensions are part of the Executive/applications interface since the Ada Run time kernel is a shared and resident part of the Executive environment. This means a need for the standardization of the highest layer of Ada RTS, which is outside the scope of this study, and the definition of services provided by the Ada extensions which are described in the next chapter. Finally, requirements for services provided to the applications and which are not directly linked to Ada features are identified: they deal with communications, health monitoring, data saving and restoring, maintenance functions and debugging facilities. Downloading kernel, memory management and the supervisor function do not require any interface services.

Requirements for Ada real-time extensions The aim of this study is to define mechanisms and services which would complete or replace the features provided by the Ada programming language and its real-time executive software.

The use of the Ada language in integrated modular avionics implies defining real-time extensions which meet avionics requirements such as time management, periodicity, task management and resource sharing. So a large description of real-time extensions has been provided making it possible to cope with the weakness of Ada.

The CIFO 3.0 proposal, sponsored by the Association of Computing Machinery, defines a solution relevant to the IMA requirements since it proposes a broad variety of services, now includes a subset of the ExTRA services and is involved with the Ada 9X process.

However it does not provide any implementation of its services so far. The ExTRA group, sponsored by the French Department of Defence, also defines a solution to the IMA problems and provides a model of implementation in Ada.

However it may not be possible to accept CIFO and ExTRA services since problems of certification with Ada tasking can occur.

In the expectation of appropriate implementations of CIFO/ExTRA services, a more relevant solution for the IMA requirements would be to use OTR-like services, which solve the major Ada Real-Time problems (Time control, Cyclicity, Asynchronous task management, Resource sharing . . .) plus Fault Tolerance, dispatching and certification mechanisms.

Ada specification of APEX services   An international working group, i.e. ARINC 653 deals with the application/executive interface. The aim of this group is to provide the absolute minimum list of services required by IMA systems using the Ada programming language for their specifications without any specific Ada feature.

For the IMAGES project the Ada programming language is used to develop the application software. Other important topics must be taken into consideration, namely that the Ada run time system belongs to the Executive Software, and that the applications can be set in any core module. One objective of the IMAGES project is to define an interface between the Executive Software and the applications in order to keep applications as independent as possible of the Executive Software.

One of the major issues of the Ada language is that it contains an internal run time system (RTS) which is called implicitly through the Ada programming language semantics. These implicit calls are defined by the Ada compiler suppliers and are not currently standardized.

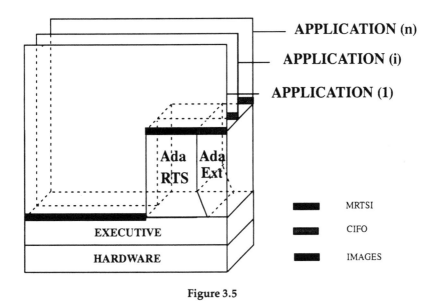

Figure 3.5

This part is not dealt with in this study because it implies the involvement of all the Ada compiler suppliers. The Ada run time environment working group (ARTEWG is an American working group) is currently defining this interface presented in the model run time system interface (MRTSI) document.

The Executive/application interfaces are defined through specifications of functional services and requirements for Ada real-time extensions in order to give the Ada run time system a deterministic behaviour. Most of these extensions are provided by the Catalogue of Interface Features and Options (CIFO) document from the ARTEWG.

The definition of the Executive/application interface is based on an Ada specification. Through this definition all the characteristics of the Ada programming language, namely the Ada exceptions, the generics and so on are used.

Results of this study must be seen as a first approach in the definition of the Executive/application interfaces. This is partially because some internal data types must be defined and standardized with agreement of the Ada compiler suppliers.

Example of use of the APEX interface This study illustrates the use of the Executive/application interface specific application, namely the control function software of the head up display computer.

Some capabilities that are included in the application today will be the responsibility of the Executive Environment Software tomorrow. This study shows a way to manage this new sharing, through the architectural design of the application software.

Some functions that were directly managed by the application are part of the IMA Executive:

- power on tests and hardware cyclic tests

- interrupt handlers

- non-volatile memory management to store failure reports

IMA Executive also provides some services to manage:

- the scheduling of the application

- the input/output of the application.

A design method was chosen in order to be able to show this new software architecture in the IMA context. The HOOD method was selected and analysed since it suits Ada, the preferred language in IMA critical embedded software, and since it is widely used in Europe.

The evolution of the HOOD method (Issues 2.2, 3.0, 3.1) was also studied, and specifically the enhancements to be expected from the latest one. The way of using HOOD in the IMA context even without Ada tasking was explained and detailed.

Special attention was paid to testability and integration for the application and the benefits and constraints brought by object-oriented methods and the Ada language.

So object-oriented and functional approaches were compared. It was shown that functional designs with associated traditional global data required a greater effort in software testing activities than object-oriented design. Object-oriented design also reduces the time spent in integration phases.

Finally, the control function software of the head up display computer was designed with the HOOD method.

The study focussed on the real-time aspects. The scheduling of the tasks is performed by the use of services provided by the Executive Environment Software. How to declare and activate periodic tasks is especially shown in the design.

The architectural design documentation shows the Executive subsets required for each part of the application. The result of this study can be considered as a mock-up that illustrated the use of IMAGES concepts and services on a specific application.

The HOOD method provides an interesting way to formalize the software design. Experience shows that both graphic diagram and textual description are not difficult to understand, even for people who are not expert in the method. The strongest and the weakest parts of the design are generally pointed out, which makes the design reviews easier.

Testability and integration are improved by proper object-oriented design, in comparison with a traditional functional approach. This will be even better if IMA Executive simulation tools are available on host computers, and if a debug kernel is provided by Executive Software on the target computer for flight tests.

It is difficult to go more deeply into the study since IMA Executive services have not been implemented yet. However some of them are very close to the services provided by the Ada connection to a hardware Executive OTR developed by one of the partners, which is operational today with non-Ada tasks. That is why the application scheduling part, for instance, is described more precisely than the other parts.

In addition, some details of IMA may change again because the concepts are not mature at the moment. Some specific problem are to be solved. For example, the sharing of responsibilities between application and Executive is to be defined more clearly in the maintenance function.

3.3 Ada language studies for avionics applications implementation

3.3.1 Introduction

The Ada programming language was chosen to develop the application software in an IMA context. However, using the Ada programming language is not without consequences. This language was designed to be used on bare machines. That is to say, it already contains a run time system (RTS) able to manage several Ada tasks and to provide an interface with the hardware resources of the target.

Ada tasking is typically non-deterministic and this problem must be resolved before any embedded use. Another problem is to know how many executive functions have to migrate to the control of the application software.

Study of the Ada in the IMAGES project deals with all the aspects of the Ada programming language that have an impact on the IMA concept. The first is to perform detailed description of the RTSs which are proposed by different Ada compiler suppliers. This step is necessary in order to interface the IMA Executive Software and commercial RTSs. Another point is to know if the Ada tasking suits well all the avionics and IMA constraints. This kind of study is also performed for the exception mechanisms, elaboration mechanisms, time management and interrupt management provided by the Ada programming language.

3.3.2 Catalogue of RTS supported features

The purpose of this study is to highlight the different features contained within an Ada run time system through the evaluation of RTS provided by two different compilers (the TLD Ada 1750A cross-compiler and the DDC International cross-compiler system DACS-80286).

The Ada run-time system may contain the following features.

- Tasking management: performs creation, activation, termination, rendez-vous, scheduling and task descriptor management.

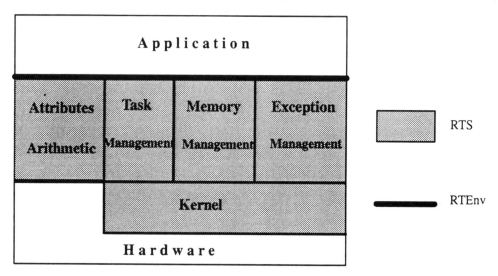

Figure 3.6

- Dynamic memory management: performs resource allocation and dealloca-tion, and can raise an exception related to the data storage when it is necessary. This feature uses two memory regions to store Ada allocated objects (stack and heap) dynamically. For allocation, different algorithms such as best-fit or first-fit strategies can be used to find chunks of space. Garbage collection may be used for deallocation.

- Interrupt management: performs interrupts handling. The Ada program-ming language supports interrupt handling via task interrupt entries. Two solutions are proposed for handling interrupts.

- Time management: supports the Ada predefined package CALENDAR and the implementation of the delay statement.

- Exception management: performs predefined and user-defined exceptions handling. Two techniques seem to be in favour for locating and selecting exception handlers: dynamic tracking and static mapping. Static mapping requires a search to locate the set of handlers for selection, whereas dynamic tracking requires a stack of exception handlers to be maintained at run time.

- Inputs/outputs management: performs low level inputs/outputs handling.

- Code sequences: includes all basic operations such as multi-world calcula-tion, block moves, string operations, Ada attribute calculations, generics, case, etc. For the case statement, jump tables are used when the alternatives are explicit, and 'if sequences' are used for scalered case alternatives. For the generics, either the code is duplicated or it is implemented by a single body with extra parameters used to distinguish between implementations.

- Other issues: elaboration, standardization (CIFO, EXTRA), etc.

The results of this sub-task are not representative enough, as only two compilers have been taken into account. Therefore, this study consists of a general description of an Ada run time system supported features.

Through the study of these two compilers, it appears that according to the Ada compiler implementors the Ada run-time system may have a different architecture, and the implementation strategy may be different from one compiler to another.

3.3.3   The Ada tasking in avionics applications

The purpose of this study is to deal with the interest of Ada tasking in an avionics context and how Ada tasking may be introduced in a multiple application architecture.

The first part of the study introduces the different features of Ada tasking in order to point out the pros and cons of Ada tasking when using it in an avionics context and presents a possible description of the Ada run-time system. An Ada run-time system is composed of a set of libraries and a kernel. Some libraries are linked to the kernel and perform 'executive' functions. Then the different features of the Ada tasking that are allowed to vary among different implementations are presented.

The second part deals with the requirements of Ada tasking in an IMA context. So the weakness of some Ada tasking features in the framework of the IMA context, which is a hard real-time context, are presented. Specific points which are discussed are: asynchronous communication, broadcasting, segregation, application monitoring, scheduling, pre-elaboration and multiple application.

Some solutions are proposed in order to solve these weaknesses.

- The passive task optimization is to eliminate certain unused threads of control for asynchronous communication, while keeping the semantics of the program unchanged.

- The rate monotonic scheduling algorithm guarantees that for a set of independent periodic tasks which ensures a sufficient condition, all the tasks of this set meet their deadlines. In order to bound the duration of priority inversion the basic priority inheritance protocol may be used, but in order to reduce their number it is preferable to use the priority ceiling protocol. When aperiodic tasks are used, two other algorithms are proposed: deferrable server algorithm and sporadic server algorithm.

- With segregation it is necessary to respect certain rules or to use a MMU to improve task segregation.

- A DELAY_UNTIL primitive may be supplied by the run-time system taking an absolute time parameter to create periodic tasks or a high-priority task which handles timer interrupts.

- An asynchronous transfer of control is possible, allowing processes to be stopped when a critical application does not succeed in meeting its deadlines, and requires actions to be taken.

- An architecture is proposed to support the multiple application concept. This model uses only one Ada run-time system and a dispatcher. The Ada RTS is in charge of scheduling all the tasks of each application and the dispatcher all the applications with their own time slices. The code of the Ada RTS is shared by all the applications and the data for the configuration of the RTS are defined for each application.

3.3.4 The Ada exceptions in avionics applications

The purpose of this study is to deal with the interest of Ada exception mechanisms in the avionics context.

The first part of this study presents and analyses the fault tolerance mechanisms. These mechanisms are placed at a high level of abstraction, independent of the chosen programming languages and development tools. Problems of detection, localization, diagnosis and perception are widely discussed.

The second part presents: in a first step, the features related to the Ada language exception mechanisms; in a second step, the use of the exception mechanisms by the application (horizontal view) specifying the correlations that exist between the exception mechanisms and the other features of the language; and in a third step the implementation of exception mechanisms (vertical view) explaining the entities that intervene during the exception mechanisms. Greater attention is paid to this last point.

The third part takes the avionics constraints into account and tries to define the usefulness of the Ada exception mechanisms in an IMA context. A set of avionics constraints is imposed by various aspects of the IMA context such as software integration, software certification, software maintenance, performance and reliability aspects. Through this study the following points are focussed on.

- How to use the exception mechanisms: for exceptional situations or to redirect program execution. Then one shows how it is possible to use exceptions in order to implement fault tolerance mechanisms described in the first part.

- How to implement the exception mechanisms: a discussion is presented about the consequences of the added constraints on the choices of implementation as far as they are concerned. Specific problems are taken into account, namely portability, maintainability, performance and reliability.

- Exceptions handling: several representative examples are presented and the conclusion of this study is to use an exception monitor, knowing the progress of the entity checked, in order to either acknowledge its malfunctioning, or diagnose the cause, or correct and resume normal functioning at the desired point.

- Propagation of the exceptions: in order to increase maintainability it is preferable not to propagate a predefined exception but an exception with a name that specifies its meaning at the functional level. Moreover the propagated exceptions should be supplied at the specification level.

- The use of exceptions for tests: this raises several problems. The first is the

commandability and the second observation. Using exceptions implies having in the programming environment a tool able to provide the exception
tree in which the exception and the exception handlers are given for each
frame.

- Extension: several extensions are proposed in order to increase the reliability
  and maintainability of the Ada programming language, namely raising an
  exception by the RTS when a deadlock occurs, eliminating imprecision in the
  predefined exceptions, raising an exception when an execution time limit is
  exceeded, adding to the accept statement an exception handler, adding the
  notion of exception type.

- Standardization: in order to be able to load several applications developed in
  different environments in the same environment, it is necessary to have a
  standardized run time system interface. Standardization of the interface
  between the application and the Executive Software which only relates the
  exception aspects is presented.

The last point presents some rules of use of the Ada exceptions in the IMA
context.

3.3.5 Complementary study

The major impact of the Ada language on the Executive is the run-time
environment (RTE) supporting the advanced features of the Ada language,
which is integrated into the Executive Software. This makes it necessary to
study how the components and features of the RTE interact with the Executive
Software. This study examines three particular features of the RTE: elaboration,
interrupts and time management.

Elaboration is described as the process by which a declaration achieves its
effect. This elaboration includes not only the initialization of data objects, but
also the computation of descriptors for types, the implementation of tasks and
generics, and the initialization of the program itself. This study first identifies
the Ada functions that cause the elaboration code to be produced and when,
within the execution of a program, elaboration can occur. Following this, the
interaction of elaboration with particular IMA constraints are examined, in
particular, how the elaboration of the RTS itself interacts with the Executive,
the potential for the application to control the order of elaboration, the needs of
the Executive when performing reset logic, and the control over elaboration
required when downloading an application onto the core module.

It is necessary for the Executive Software to have explicit control over the
initial elaboration of an application so that the reset logic can be performed
correctly, and also to permit the elaboration of an application to be performed

separately (and before) the activation of the application. This requires that the RTE be modified to allow this direct access. Moreover, this direct access has to force the application to return control to the Executive on completion of its initial elaboration sequence.

It is not expected that performance issues will be a problem for elaboration, expect when a fast reinitialization is necessary after a failure condition. In this particular case, it is suggested that the use of a purpose-built synchronous scheduler and the inclusion of enhanced pre-elaboration extensions would prevent a significant loss of downtime.

It is strongly recommended that no interrupts be handled by the applications. Within the Executive, some interrupts are used by the RTE which will require the detailed study of any proposed RTE implementation. The most common use by the RTE is for the maintenance of a real-time clock. There are three items in the RTE concerned with time management, of which the delay statement is of concern to the Executive. As the correct implementation of this feature requires a hardware-controlled timer, a requirement for high resolution delay timing can result in an unacceptable servicing overhead. An alternative is to use a programmable hardware timer for specific delay periods.

As a separate part of the study, a prototype of three modelled avionics applications supported by an Executive (containing the Ada RTE) is examined. The applications are modelled on specifications reported in the APEX interface requirements and are essentially fully deterministic control loop applications. The Executive is built around a commercial multi-application kernel supplied by TLD as an integral part of their Ada compiler system. This prototype exercise is developed to the identification of the Executive interface and functionality.

It is found that some problematic areas exist, particularly when redundant application lanes are executed in separate cabinets. It is not yet clear how these applications are to communicate with each other, notably for the purpose of identifying the master lane. A solution, used for this exercise, is to broadcast a message to all cabinets and require that each core module within each cabinet determines if this message is applicable.

A further problematic issue is the difficulty in developing an Executive that can handle any application. Whereas it is possible to create a general purpose Executive to handle scheduling, memory management and other such functions, it is not so clear how the disparate I/O needs, the differing error response requirements and the multitude of entries in the data base can be handled without customizing each Executive to a particular set of applications.

3.4 IMA impact on the development and certification process

The issues of certification and development process began to be studied during this project.

3.4.1   Certification issues

This study describes some V&V techniques in use today and analyses the possible changes required to certify software (Executive Software and application) in the IMA context. The critical aspects of the software certification identified, and the differences compared with the actual certification guidelines DO-178-A/B are emphasized.

Therefore, this study analyses the Executive Software and the applications software in an IMA cabinet, evaluating the possibility to maintain independence between the applications and the cabinet hardware. So it would ensure that an application which is certified by the airworthiness authority can be loaded in the cabinet without recertifying the whole cabinet, but only validating the main integration. This is the goal of the IMA software architecture. For this reason it is possible to study the application certification and the Executive Software certification separately. The software criticality level for each software module are analysed and described. Current V&V techniques are discussed and evaluated for applicability in the IMA context. The incremental techniques that proved to be applicable for IMA software development are also described.

Finally, the study present the Italian Airworthiness Authority's (RAI) point of view and some proposals for further studies; for example, the implementation of standard certification tools for an automatic testing method.

3.4.2   Integration issues

The purpose of this study is to study the integration of the applications within any IMA cabinet. So it is supposed that Executive has been integrated on HW module and validated previously.

The main assumptions about the Executive have led to the particular integration way proposed in this study. The most important ones are as follows.

- The Executive provides robust partitioning between the embedded applications.

- The Executive is qualified to run any set of applications as long as they remain within its domain of use.

- The integration of application software in IMA context is fully compatible with DO178A/B requirements. Remarks deal only with the development organization.

Hence, a life cycle and an organization are proposed. Deduced from the full-IMA concept, application software developments are perfectly separated from the LRMs (and Executive) developments. So, it appears that a new activity is required (compared with classical LRU development); it deals with

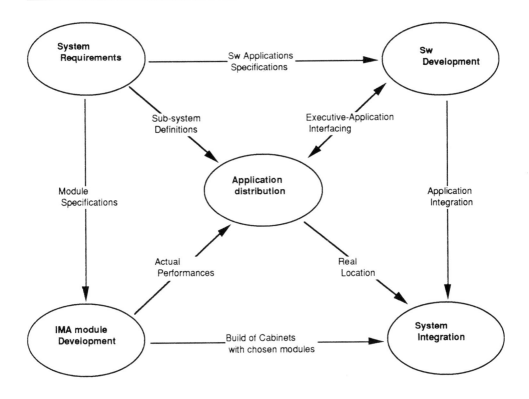

Figure 3.7   IMA life-cycle.

the management of the distribution of the applications within the cabinets. Symmetrically, application integration also includes cabinet integration.

The integration of one application must be considered as a software/software integration disconnected from the hardware layer. By merging one application code with the Executive, this integration stage deals with checking that the application meets the requirements of the real Executive Software but not with validating the application (this must have been done before by the software supplier). A connected issue is how to prove the compliance of the development environment used by the software supplier with the actual Executive.

Associated with this stage, the document describes, among other features, the errors to be detected and the content of the deliveries. Then special attention is paid to the needs for standardization and for identification, and to the acceptance tests.

In a similar manner, the multiple applications integration is studied. For convenience, it is split into two steps: core module integration and cabinet integration.

The purpose of such stages is firstly to check the consistency between Executive environment and applications, and secondly to prepare the loadable tables which will configure the Executive. In no way is incremental integration of the application required because it is assumed that the Executive performs perfect segregation between the embedded applications.

Dealing with the same principle, cabinet integration must be distinguished from functional avionic validations because any avionic function extends over more than one cabinet.

Finally, particular interest must be paid to the software quality and configuration management as each cabinet receives numerous applications.

3.5 Studies of communication between applications

3.5.1 Introduction

In the IMA concept the different applications are dispatched among several cabinets and each cabinet may contain one or more core processing modules. Within such an architecture, buses (ARINC 629 or ARINC 659) are used as standardized communications media between the modules.

Communication and data flow between applications take place along many separate paths within an IMA architecture. This communication must be handled by the executive and must be transparent to the applications.

3.5.2 Communication requirements

The different topics covered by the study on the data communication requirements of an IMA system are as follows.

The relevance of the OSI model to IMA architectures: the Open Systems Interconnection (OSI) model for network protocols is an International Standard for the structure of communications over a databus network. It consists of seven layers, each of which handles different functions necessary for communications to take place. One problem with the OSI model, however, is that it does not address real-time issues. This means that some of the layer functions, especially in the session layer, are not relevant to operation in an IMA environment. A revision to the OSI model is proposed, therefore, in which two of the layers could be removed.

The effect of reconfiguration on the transport layer: the transport layer is responsible for the logical routing of messages between applications. If recon-

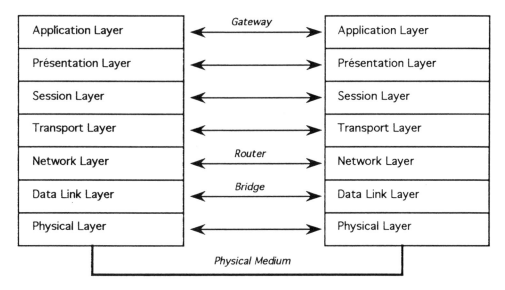

Figure 3.8

figuration of applications occurs, the transport layer must be able to change the routes of messages to the new location of reconfigured software. It is assumed that multiple copies of applications will be executed as hot standby applications, which take over if the primary copy suffers a failure. The transport layer must be able to transmit messages over a variety of different paths using a variety of equipment such as bus bridge modules and I/O modules. The information available to the transport layer about the location of applications must be updated by the RTE kernel if reconfiguration occurs.

The effect of redundant buses on the network layer: it is probable that an IMA cabinet will have multiple ARINC 659 backplane databuses for hardware redundancy reasons. If this is so, then the network layer of the protocol model will have to handle reconfiguration of these buses if one of them fails. This could be done through a voting system, or a master/slave system. All the CORE modules in an IMA Cabinet must agree that a particular bus has failed, and switch it out of service. Mechanisms for doing this must be developed.

Aspects of communications latency and scheduling: there will always be some delay in the flow of data between sensors/actuators and the applications in core modules, and this delay must be taken into account by applications. In order to make the data available to different core modules as consistent as possible, the use of a distributed data base is proposed, which is constantly updated as new data are made available. This can ensure that it is consistent with other data bases within a given time tolerance.

Error detection: errors in the transmission of data over a databus will always exist. However, their probablity of failure, and hence their average rate of occurrence, can be determined. Once way of detecting a failure on a databus is to measure the error rate of a bus, and compare it with the expected error rates for normal operation. If the measured rate is much higher than the expected rate, then it can be assumed that the bus has failed. There must be some software which monitors the errors, and the protocol must report errors to this software.

Characteristics of applications studied in the BRITE/EURAM 'Optical Data Transmission' project. A few systems currently in service on aircraft were studied, to obtain an approximate estimate of what types of data were used, and how much data might be transmitted over an IMA system.

3.5.3 Communications between applications

This study is split into two parts.

The first part studies two representative and opposite modes of communication between applications and their impact on the Executive Environment Software. This is based on the fact that two of the main questions about communication issues are: what are the communication means usable by the application program?, and what is the relationship between the execution of one application and actual I/O? Two ways of answering give two representative cases of communication philosophies.

The first case renders the communications issues transparent for the applications and assumes that the Executive schedules inputs/computations/outputs for each of them.

So the Executive has all the knowledge concerning the application needs: when all its input data are available, when all its computations are over and then when its outputs can be made. In fact, the basic architecture deals with a sophisticated and global memory space managed by the Executive according to real-time constraints and access rights.

As soon as it is assumed that the backplane bus is synchronous, it becomes obvious that the application computation is under the control of the communication features.

Several interesting characteristics can be listed:

- communications are fully deterministic

- data transmission is similar to accessing shared memory

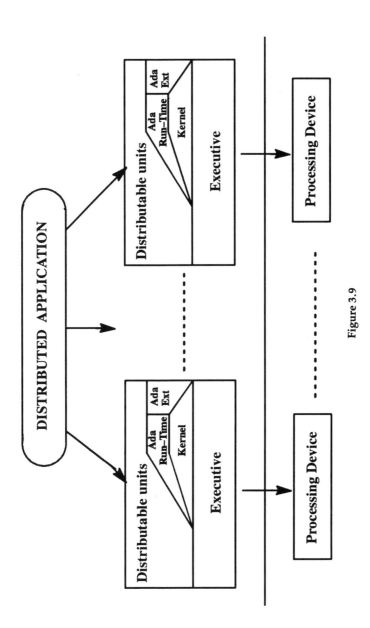

Figure 3.9

- in such a context, 'application software' is actually a set of partitions of the present application software

- as data transmission controls the execution of the applications, the design stage must give an order to the application scheduling

- parallel processing within the core module is forbidden (within the cabinet, it is difficult to provide)

- the backplane communication frame depends on the period of the least frequent partition

- in fact, everything needs to be frozen at the design stage.

The second case supposes that applications call Executive services for all communication issues and that physical transmissions are managed at the level of the whole module. So the relationship between real-time execution and communication must be defined. For instance, as execution windows can be given to an application, a window can be addressed for communication issues at each module cycle. The basic architecture, from the point of view of the applications, deals with the use of a dynamic data base which manages the data updating and backplane bus transmissions.

Several interesting characteristics can be listed:

- communications between modules can remain asynchronous and based on periodical refreshing

- as applications control the piece of data to be transmitted, overheads can occur on the backplane bus

- access conflicts to the backplane bus can also occur and input information can be old

- information to be transmitted on the asynchronous backplane bus must be included in messages

- computations within the module look like parallel ones

- each module communication frame is linked to the most frequent application and by principle, independent of the other modules.

The second part of the study compares these two modes from the points of view of the configuration of the Executive and of the impact of modifications. These modifications deal with upgrading of the application code and moving of their executive location.

In a first step, configuration aspects are studied. For the first case, it appears that communication configuration extend over all the cabinet items: it is true for any partition real-time aspect as well as location of each piece of data in the memory. Some considerations are given to the Ada implementation and the link between the language and hardware memory management facilities.

For the second case, configuration deals with programming the dynamic data base for data routing, message building and memory space reservation.

In the second step, the impact of different kinds of modifications is considered in detail. With the first communication philosophy, the configuration of the whole cabinet is modified in most of the cases. With the second one, the impact can be more local as requests to the data base use symbolic labels.

3.6   Distributed system architecture

3.6.1   Requirements for distributed computing

A distributed IMA hardware architecture can be generalized to be composed of a set of heterogeneous processing devices making out the nodes of the platform. The corresponding software architecture on each node is identified to consist of an application part with an additional Ada run-time system, kernel and Ada extensions, together with an executive part. Furthermore, an IMA distributable application is then considered to be composed of various of distributable units expressed in the domain of Ada. The total distributed architecture is illustrated by Figure 9.

Several topics was discussed through this study.

Ada and distribution: the involved partner describes the requirements from the application point of view. That is, the requirements arising from the intended use of Ada on an IMA distributed platform are analysed and put forward. This study presents a characterization of the integrated modular avionics (IMA) distributed platform and an exploration of the problems and approaches taken when Ada is to be used on such platforms as a distributed programming language. All approaches are analysed and settled through formal models following the school of algebraic specifications. In this way, a sound and consistent taxonomy is obtained, against which trade-offs and decisions regarding Ada and distributed systems can be made reliability and un-ambiguously.

Distributed operating systems: the involved partner describes the executive requirements and procedures from the operational point of view for distributed computing across multiprocessor systems. User transparency features end an extension of the mechanisms related to scheduling and communication. This study presents a survey on distributed operating systems. As an introduction the basic requirements for distributed operating systems are presented to the reader. The capability of integrating a distributed operating system into the IMAGES concept is examined with special regard to communication aspects and scheduling strategies. Investigations on these operating systems have been made with respect to real-time capabilities and HW independence.

Report on requirements for resource allocation in real-time distributed systems: the involved partner generates a pragmatic definition of existing systems describing the requirements for resource allocation independence of the present load system and in case of failure, i.e. reconfiguration of the system. This study proposes a list of requirements which will be considered within the development of the executive software. This requirement list is elaborated after evaluating some of the impacts of the many choices available. This covers the analysis of currently available IMAGES documentation at the status of draft issues. The requirements list will consider the following three topics:

- resource location

- accounting and resource control

- fault tolerance and reconfiguration

3.6.2   Impact of different architectures on Executive Software

The purpose of this study is to compare distributed and centralized architectures and state the requirements for an Executive for both. The architectures cited are those proposed by BAe and BOEING and described in Arinc Standard 651 'Design Guidance For Integrated Modular Avionics'.
   The features of the architectures are described using information obtained from ARINC 651. The components of the architectures are the same and consist of the following:

- core modules

- I/O modules

- ARINC 659 backplane databus

- ARINC 629 airframe databus

- remote data concentrators

- smart sensors and actuators

It is the implementation of these components that distinguishes the centralized architecture of BOEING with only one locus of control, from the distributed architecture of BAe in which the nodes cooperate to provide a hierarchy of control. The use of remote data concentrators removes the task of data processing from the applications in the cabinets. This, and the minimal use of dedicated I/O modules with the corresponding increase in reliance on databusses and smart actuators/sensors provide the flexibility of a distributed system to the BAe architecture.

Different approaches to fault tolerance are apparent. BAe allows dynamic task reconfiguration and a distributed executive and BOEING does not.

The interfaces that would bound the IMAGES Executive are then discussed, namely the APEX COEX and HMEX (as defined in ARINC 651). The definition of these is of great importance if the Executive is to be truly portable across distributed and centralized computing platforms.

The Executive is discussed function by function using the breakdown detailed in the IMAGES project. From this it can be noted that an executive for both types of platform requires all the functional blocks described as the goal of the IMAGES project is to define a Generalized Executive. The implementation of the functions will of course be constrained by the architecture, but the requirements will remain the same. One important thing to note is that a truly distributed Executive would not be used as communication overheads would be too high.

4 Conclusion

The IMAGES project focussed on the definition of the Executive Software of the integrated modular avionics context.

In regard to the Executive Software, all the services were defined, namely the real-time Executive features, communication management, health monitoring function, downloading and debugging kernels, memory management, initialization, interrupt and exception handling, and the avionics components library.

In order to control the workstate of the whole cabinet in the aircraft and the management of reconfiguration, a function named supervisor was analysed.

In the IMA concept, the Executive Software must be able to manage multiple applications written in Ada. Each application could have several tasks. In the IMAGES project, the Ada run-time system has been chosen to be resident within the Executive Software and shared by all the applications.

In order to have portable applications, a standardized application/executive interface was defined. This interface provides three different sections:

- operational services

- standardization of the Ada RTS interface used by the Ada compiler to provide the executable code of the applications

- Ada real-time extensions required by the avionics applications.

The second point was raised but solutions depend on agreement between Ada compiler suppliers more than between aeronautical industries.

In regard to the development process, the study of the application integration highlights the need for a new life cycle. Thus, first contacts with airworthiness authorities were made to initiate the new process of certification in the IMA context.

During the project, it often appeared that it was necessary to engage other studies on topics concerned by the IMA concept, in order to make such a concept mature.

In regard to the Executive Software, the next step consists in its implementation which will require making several main choices of which the nature of the backplane bus is far from being unessential.

In regard to the Ada programming language several points were highlighted. The first is that in order to be able to use any Ada compiler for the application development and keep the possibility to run this application on a RTS different from the one used without the Ada compiler, it is necessary to define a standard interface for the RTS. This problem could be solved only if all the Ada compiler suppliers agree. The second point is that, as IMAGES supposes that the Ada RTS is included within the Executive Software, further studies must be conducted in order to define exactly how these two software parts can live together. Another concern is to evaluate the introduction of specifications in Ada programming to increase fault avoidance, fault removal and fault tolerance, to know how to define extensions of Ada features to facilitate the programming of dependable software, how to couple Ada compiler and the memory management unit, and to address the introduction of new tolerance techniques in a multi-tasking context (monitor task), in order to have more influence on the standardization of the Ada environments.

About standardization, the problem is to know how far aircraft manufacturers want to go. Having a standardized hardware support or a standardized interface between hardware and software. According to these standardizations, what is the impact on the Executive software?

About the Supervisor function, its reliability is a subject which will take some more study. Some items of this study are:

- supervisor (self) monitoring and redundancy management

- hierarchy of and communication between supervisor processes
- supervisor fault handling (activation/deactivation of a supervisor process)
- supervisor core module election algorithm
- supervisor and maintenance function interface.

Index

Index compiled by Sheila Shephard